France
Since 1945

Robert Gildea

Oxford   New York
OXFORD UNIVERSITY PRESS
1996

Oxford University Press, Walton Street, Oxford OX2 6DP

Oxford New York
Athens Auckland Bangkok Bombay
Calcutta Cape Town Dar es Salaam Delhi
Florence Hong Kong Istanbul Karachi
Kuala Lumpur Madras Madrid Melbourne
Mexico City Nairobi Paris Singapore
Taipei Tokyo Toronto
and associated companies in
Berlin Ibadan

Oxford is a trade mark of Oxford University Press

British Library Cataloguing in Publication Data
Data available

Library of Congress Cataloging in Publication Data
Data available

ISBN 0–19–219246–9

1 3 5 7 9 10 8 6 4 2

Typeset by Pure Tech India Ltd., Pondicherry
Printed in Great Britain
on acid-free paper by
Bookcraft (Bath) Ltd
Midsomer Norton, Avon

To my Father

DENIS GILDEA

who worked in the British Civil Service
1948–83
towards European union

Acknowledgements

I am indebted in the first instance to Merton College, Oxford, and to the Discretionary Fund of the Regius Professor of Modern History of the University of Oxford, who kindly financed a research trip to Paris in 1993. I would like to thank the staff of the libraries I have used, notably those of the Bibliothèque Nationale in Paris, the Bibliothèque de Documentation Internationale Contemporaine at Nanterre, the Bodleian Library, Oxford, and the Maison Française, Oxford. I am particularly grateful to the Librarian of the Maison Française, Ms Rosenchild-Paulin, for her friendly and unstinting assistance. A first draft of the text was read by Sudhir Hazareesingh of Balliol College, Oxford, and by Michel Boyer of the Institut d'Études Politiques of Lyon; their suggestions and corrections have been invaluable. I would like to thank my editors at Oxford University Press, Catherine Clarke and George Miller, for their advice at different stages of the writing of this book. Lastly, I want to thank Lucy-Jean, Rachel and Georgia for their support and patience, and for helping me to complete the first draft before William was born.

Contents

Figures and Tables

Abbreviations

AIS	Armée Islamique du Salut
ARC	Action Régionaliste Corse
ARS	Action Républicaine et Sociale
CAP	Common Agricultural Policy
CD	Centre Démocrate
CDP	Centre Démocratie et Progrès
CERES	Centre d'Études de Recherche et d'Éducation Socialiste
CFDT	Confédération Française Démocratique du Travail
CGT	Confédération Générale du Travail
CIR	Convention des Institutions Républicaines
CNIP	Conseil National des Indépendants et Paysans
CNJA	Cercle National des Jeunes Agriculteurs
CODENE	Comité pour le Désarmement Nucléaire en Europe
CODER	Commission de Développement Économique et Régional
EDC	European Defence Community
EMS	European Monetary System
ENA	École Nationale d'Administration
ESSEC	École Supérieure des Sciences Économiques et Commerciales
FAR	Force d'Action Rapide
FFI	Forces Françaises de l'Intérieur
FGDS	Fédération de la Gauche Démocratique et Socialiste
FIS	Front Islamique du Salut
FLB	Front de Libération de la Bretagne
FLN	Front de Libération Nationale
FLNC	Front de Libération Nationale Corse
FLNKS	Front de Libération Nationale Kanake et Socialiste
FNSEA	Fédération Nationale des Syndicats d'Exploitants Agricoles
FRC	Front Régionaliste Corse
F.-T.P.	Francs-Tireurs et Partisans
GATT	General Agreement on Tariffs and Trade
GIA	Groupment Islamique Armé
GRECE	Groupement de Recherche et d'Études pour la Civilisation Européenne
HEC	Hautes Études Commerciales
HLM	Habitation à Loyer Modéré
INSEAD	Institut Européen d'Administration des Affaires

INSEE	Institut National de la Statistique et des Études Économiques
IUT	Institut Universitaire de Technologie
JAC	Jeunesse Agricole Chrétienne
MDF	Mouvement Démocratique Féminin
MLF	Mouvement de Libération des Femmes
MOB	Mouvement pour l'Organisation de la Bretagne
MOI	Main d'Œuvre Immigrée
MRG	Mouvement des Radicaux de Gauche
MRP	Mouvement Républicain Populaire
OAS	Organisation de l'Armée Secrète
ORTF	Office de Radiodiffusion-Télévision Française
PR	Parti Républicain
PS	Parti Socialiste
PSA	Parti Socialiste Autonome
PSU	Parti Socialiste Unifié
RER	Réseau Express Régional
RMI	Revenu Minimum d'Insertion
RPCR	Rassemblement pour la Calédonie dans la République
RPF	Rassemblement du Peuple Français
RPR	Rassemblement pour la République
SFIO	Section Française de l'Internationale Ouvrière
SMIC	Salaire Minimum Interprofessionnel de Croissance
SMIG	Salaire Minimum Interprofessionnel Garanti
SONACOTRA	Société Nationale de Construction de Logements pour les Travailleurs Algériens
UDB	Union Démocratique Bretonne
UDF	Union pour la Démocratie Française
UDR	Union des Démocrates pour la République
UDSR	Union Démocratique et Socialiste de la Résistance
UDV^e	Union des Démocrates pour la Ve République
UPC	Union du Peuple Corse
ZUP	Zone d'Urbaniser à Priorité

Introduction

This is a concise study of French national identity, culture, obsessions, and aspirations in the fifty years after the end of the Second World War. The starting-point of 1945 is no doubt of less significance in French history than 1944, the year of the Liberation of France from German Occupation, or 1946, date of the constitution of the Fourth Republic. Neither does that fifty years form a bloc: it has seen two Republics, a period of economic growth followed by one of economic stagnation, the recovery of great-power status but also a loss of national confidence. The end of President Mitterrand's term in office, however, after fourteen years, echoing in some way the eleven-year presidency of de Gaulle, provides a caesura and moment for reflection. It may also, of course, be argued that these have not been the fifty most exciting years of French history: no European wars, no revolutions. Wars of course there have been, notably the Algerian war of 1954–62, while the events of May–June 1968 may be considered France's last great revolution. But it has been a period of immense challenges for the French: constructing a new European order, building a modern economy, searching for a stable political system. It has also been one of anxiety and doubt. The French have had to come to terms with the legacy of the German Occupation, with the loss of Empire, with the influx of foreign immigrants, with the rise of Islam, with the destruction of traditional rural life, with the threat of Anglo-American culture to French language and civilization. All along there has been a battle to meet the challenges of the late twentieth century while preserving the historic characteristics of French identity.

This book is divided into eight chapters. The first two and last two examine the French political system and France's role in the world, and split the fifty-year period at two different points. Chapter 1 looks at France's attempt to recover national greatness after the Second World War, its ambivalent relationship with American imperialism, its attempt to deal with the fear of German resurgence by building the European Community, and its struggle—eventually doomed—to preserve its Empire. It ends with the conclusion of the Algerian war in 1962 and considers the legacy of that war. Chapter 8, mirroring it in some way, examines the way in which the foundations of contemporary French foreign policy were laid by de Gaulle. These included the carving out

FIG.1. France: Departments and Regions.

Source: *L'Etat de la France 94–95* (Paris, Éditions La Découverte, 1994).

of an independent role for France in a world dominated by the superpowers, ensuring that French hegemony prevailed in the European Community, and the development of a neo-colonialism to preserve its influence in Africa and the Pacific. It also, however, traces the frustration of these dreams: the loss of French bargaining power *vis-à-vis* the United States after the collapse of the Soviet Union, the displacement of France by Germany as the dominant power in Europe after the reunification of the latter, the souring of its neo-colonial ambitions, and the attempt to build an alternative forum of influence in the French-speaking world.

The second and seventh chapters examine the evolution of the French political system, and are divided by the fall of de Gaulle in 1969. Chapter 2 deals with the re-establishment of the Republic, which had been abolished in 1940, and the construction of the Fourth Republic as a parliamentary Republic in the teeth of challenges from Communist revolution and Gaullist dictatorship. It endeavours to demonstrate the strengths as well as the weaknesses of the much-maligned Fourth Republic. It then examines the return of de Gaulle to power in 1958, and the construction of the Fifth Republic as a presidential republic. Finally it looks at the increasingly dictatorial pretensions of de Gaulle, the recovery of the opposition, the revolution of 1968, and the crisis of state that ensued. Chapter 7 focuses on the resolution of the eternal French crisis of state by means of a pluralist democracy, allowing for the peaceful alternation of Right and Left in power, and sometimes their cohabitation in power, the growth of a national consensus, the decline of ideology, and competition between parties to dominate the centre ground. It then examines the reverse side of these developments: frustration with what was seen as a closed and corrupt political system, the growth of new parties at the extremes of the political spectrum, political disillusionment, and the emergence of a new kind of politics and politician.

The four central chapters are not divided chronologically, and each deals with a single theme. Chapter 3 looks at the shadow of the Second World War and the Occupation hanging over French life, and at attempts of the French both to come to terms with it and deny responsibility for it. It examines the construction of the myth of the Resistance, which made out that all French people were heroes, and the disintegration of that myth, as painful truths came to light of French involvement in anti-Semitic persecution and the Holocaust. It concludes with the trials of those accused of crimes against humanity committed during the Occupation and the controversy over President Mitterrand's past under the Vichy regime.

Chapter 4, more positively, looks at the post-war economic miracle that transformed France from a 'stalemate society' based on small farms and small businesses into a modern, competitive, urban and industrial system, and at the revolutionary role of economic planning. It examines the social consequences of these economic changes: the decline of the peasantry, the transformation of the working class, and the emergence of a new, salaried middle class. The chapter also, however, deals with the collapse of economic growth after the oil crisis of 1973, the struggle of governments to find a policy to deal with economic crisis, the progressive de-industrialization of the economy and de-urbanization of society, and the growth of unemployment and poverty.

Chapter 5 centres on the construction of a French national identity and the challenges presented to that project. Its first section looks at the democratization of the French education system designed to forge all citizens in the same mould, but also at the cult of meritocracy and the selfconscious cultivation of an élite designed to run the centralized French state and modern economy. The second section examines the struggle of women for political rights, equal opportunities at work, and the right to control their own bodies. But it also examines the survival of a male-dominated society and the failure of feminism to develop strongly even among French women themselves. The third section examines the political and administrative centralization of France, the growth of regionalist opposition to the centralized state, and limited measures of decentralization taken to deal with them. The final section examines the obsession of the French with a single national identity, which required minorities to assimilate French language and culture as a condition of the exercise of political rights, and to practise their religion only as a private concern. It examines the challenge to this ideology, most notably from North African immigrants and the rise of Islam, and the dilemma faced by the French as to whether to defend or redefine their identity.

Chapter 6, lastly, deals with the question of French culture. It looks at the rise and fall of that peculiarly French herald of culture, the intellectual. It examines the growth of mass culture, which tended to be synonymous with American culture, and at the threat presented by it to French culture. It look at the concern of the state to implement a policy to defend and develop French culture, and at the various strategies adopted. Special attention is give to tensions which run through many of the other chapters, between traditional and modern worlds, élite and mass culture, things cosmopolitan and things French.

1

Crisis of Empire

In search of lost greatness

In 1945 France was a great power that had come within an ace of extinction. In 1940 she had suffered the worst defeat in her history, overwhelmed within the space of six weeks. She had been occupied by the Germans (and in small part by the Italians) for four years, the so-called unoccupied zone in the south itself invaded in November 1942. Despite the internal Resistance and combats of the Free French she was liberated only with the help of the Allies, and was lucky to escape an Allied military administration of the kind that was imposed on Germany.

Defeat was compounded by the time the Allies took to recognize de Gaulle's provisional government as the legitimate government of France. The United States in particular had long hoped that the Vichy government would at some point end its policy of collaboration with Germany and swing onto their side. Not until after the Liberation of Paris in August 1944 was the French Committee of National Liberation officially recognized by the USA, Great Britain, and the Soviet Union, and France admitted to the Security Council of the newly formed United Nations along with these three powers and China. Even then, France was not admitted to the secret talks of the Allies about the post-war settlement. 'To rebuild our power: that is what is henceforth the great cause of France,' de Gaulle told the provisional Consultative Assembly in Algiers in November 1944, before flying to Moscow, to seek a counterweight to Anglo-Saxon hegemony by renewing the Franco-Russian link that had served France so well since the time of Napoleon.[1] A treaty of alliance and mutual assistance was signed between France and the Soviet Union the following month, by which each agreed that neither would make a separate peace with Germany before the end of the war, but otherwise of little real import. France was not invited to Yalta in February 1945—a fact that subsequently became an obsession of de Gaulle's—although France was granted a zone of occupation in Germany and a place on the Interallied Control Commission. De Gaulle repaid the snub by refusing an invitation from Roosevelt to meet him abroad his warship at Algiers on his way home.

For de Gaulle, the way to restoring the honour and greatness of France lay through participation in the final defeat of Germany. Even on this score the Allies, and in particular the United States, made it difficult for France to achieve any military glory. The provisional government had no military autonomy, and was refused permission to raise troops in any significant numbers until the end of 1944. French forces under General Leclerc retook the symbolically important city of Strasburg in November 1944. The following month, in response to the German counter-offensive in the Battle of the Bulge, Eisenhower, as supreme Allied commander, ordered the French to retreat behind the Vosges. The French refused, partly because of the threat of German reprisals in Strasburg but mainly on the point of honour that promises had been made to recapture and defend Strasburg and that in the annals of French history since 1870 possession of Alsace and its capital was the gauge of whether France or Germany was the dominant power. The German assault was withstood, and in March 1945 French forces crossed the Rhine into Germany, occupying Karlsruhe and Stuttgart. It was de Gaulle's intention to impose himself at the feast of the occupation of Germany, both in Baden-Württemberg and on the Rhine. But the British refused to surrender their right to Cologne and the Rhineland, the Americans insisted on their right to Karlsruhe and Stuttgart, and in July 1945 France was left with a zone of occupation that included only the Saar and bits of the Rhineland and Baden-Württemberg that the British and Americans did not want.

A love/hate relationship with the United States

After honour and greatness, France's ambition was to recover her rank among the great powers. Following the Second World War, however, she was so weakened and ruined that she required the help of the United States both for military security and for economic reconstruction. This resulted in an ambivalent attitude towards the USA: on the one hand, anxiety that the protection and benefaction might not be forthcoming, on the other, irritation at the strings attached and loss of independence involved. The overall result was a kind of petulant ingratitude.

De Gaulle had plenty of reason to feel prickly towards the Americans, but there was little change of attitude after he left power in January 1946 and foreign affairs became a fief of the Christian democratic Mouvement Républicain Populaire (Popular Republican Movement or MRP), and, in the first instance, of Georges Bidault, who had headed the National Resistance Council during the war. France and Great Britain concluded a defensive treaty at Dunkirk in March 1947, which Bidault was adamant should be directed against Germany rather than the Soviet Union. After the Communist coup in Prague in February 1948, the alliance was extended by the Brussels pact to the Benelux countries and directed against any aggressor, including the Soviet Union. But Bidault knew that such a regional pact would be of no use without the firmest possible American commitment to European defence, and pressed them for military aid and for more troops to be sent to Europe. In the summer of 1948 France had only 600,000 troops under arms, and envisaged having to abandon North Africa and concentrate on the Rhine should war break out with the Soviet Union over Berlin. France had no alternative but to join the NATO alliance in April 1949, accepting that the Atlantic treaty extended to Algeria if not to Morocco and Tunisia, but grateful to obtain American bases and American troops on French soil.

Although the dominant view in the French political class was that France could not do without American protection, there was nevertheless fierce resentment in many quarters of American hegemony. A movement among French intellectuals, often of Christian democratic persuasion, refused subservience both to Soviet totalitarianism and to American capitalist imperialism. *Le Monde*, founded at the Liberation as a paper independent of political parties and edited by Hubert Beuve-Méry, supported a third course, that of an armed and neutral Europe between the USA and USSR. One of its columnists, the medieval historian Étienne Gilson, was so forthright in his attacks on American imperialism that he was branded a traitor, forced to resigned his academic job in Paris, and expelled by the Académie Française. The Communist party denounced the colonization of France by the United States and the threat of war posed by American nuclear power, while portraying the Soviet Union as an eminently peaceable country. Excluded from power in May 1947, it sought to increase its influence by organizing a peace movement around the Stockholm Appeal against nuclear weapons in 1950. The propaganda began to bite as 'US Go Home' graffiti expressing hostility to American forces in France multiplied, and the arrival of General Ridgway, the new NATO commander in Europe, in May 1952 was greeted by riots. Communists dubbed him a war criminal and the 'microbe general' for ordering the use of biological weapons in the Korean war. Intellectuals such as Jean-Paul

Sartre and Simone de Beauvoir, who had toured the United States
between 1945 and 1947 and had not been unduly critical, were angered
by the arrest of the leading Communist Jacques Duclos during the riots,
when pigeons alleged to be carriers of Soviet messages were discovered
in his car. This insensitivity was compounded by the execution of the
Rosenbergs in the United States as Soviet spies, which provoked Sartre
to denounce the Americans as 'mad dogs' and to become, increasingly,
a fellow-traveller of the Communist party.

The French depended on the Americans after the war not only for
military security but also for economic aid. Compared with 13 departments devastated by war in 1918, 74 were devastated in 1945. Industrial
production in 1945 was 38 per cent of what it had been in 1938. France
needed to import capital goods in order to reconstruct the economy,
but also needed credit to pay for them. The United States was in a
position to provide that credit, but insisted on exacting certain conditions. The first was that the French put their economic house in order,
notably by balancing the budget and keeping inflation under control,
in order that American credits should keep their value. The second was
that the French should accede to free trade, so that the United States
could sustain its own growth by uninhibited exports, and that they
should allow the Americans access to strategic materials in their
colonies.

American aid was in no sense quick to materialize, and the conditions
were felt to be very harsh. Early in 1946, the veteran socialist politician
Léon Blum was sent to Washington to negotiate a deal with Secretary
of State Byrnes. The French requested $3 billion, but under the
agreement of May 1946 they secured only $650 million, dressed up by
various means for the French public as $2 billion. What was most
remembered about the agreement, however, was France's caving in
before the Hollywood film industry, accepting that French cinemas
would show French films for no more than thirteen weeks a year. In a
market that could absorb about 175 films a year, French studios made
only 40 films in the first third of 1947 while 182 American films were
authorized for dubbing. The relative share of the French market swung
from 45 per cent French and 40 per cent American in 1944–7 to 25 per
cent French and 50 per cent American in 1947–50.

American aid was put on a firmer footing under the plan outlined by
Secretary of State Marshall in June 1947. The presence of Communists
in the French government since 1944 had long been an obstacle to
American generosity, and, as Bidault recognized in September 1947,
'the exclusion of the Communists from the Italian government, and the

formation in France and Belgium of centrist cabinets, were the political conditions of American aid'.[2] The Americans did not stop at twisting the arm of the French government. A leader of the American Federation of Labour, Irving Brown, was sent to France by the State Department in 1945 to undermine the Communist-controlled Confédération Générale du Travail (General Confederation of Labour, or CGT). When the CGT, in protest against the Communists' exclusion from power, launched a massive strike in November 1947, Brown and the US ambassador, Caffery, co-ordinated efforts to break the strike and encouraged the splitting away from the CGT of an anti-Communist federation, Force Ouvrière.

The Americans insisted on the right political conditions for aid; they also demanded the right economic conditions. These were imposed by a series of missions in each European country receiving Marshall Aid, responsible to the Economic Co-operation Administration, based in Paris, and bilateral agreements made with each recipient power. That with France was signed in June 1948, and the three brief ministries in power between 1948 and 1949 all pursued policies of economic austerity, balancing the budget by spending cuts and tax rises, and price and wage controls to bring down inflation. The Americans also required that all barriers to their exports and investment be removed, so France was inundated not only by American products but also by propaganda selling the American way of life. 'Will France become an American colony?' asked one book in 1948, exposing the threat from American westerns and gangster movies, children's comics such as *Donald*, *Tarzan*, and *Zorro*, and magazines controlled by American trusts, notably *Reader's Digest*, called *Sélection* in France.[3]

The French won a minor victory in September 1948, when the French boxer Marcel Cerdan became world champion by beating an American in Jersey City. The real battle, however, was fought over Coca-Cola. Fed to GIs during the war, it was then the object of a sustained campaign to penetrate European markets. Coca-Cola was not simply a product, it was an image: that of the consumer society, on the wings of mass advertising, 'the essence of capitalism' in every bottle for its president, James Farley, a weapon in the global ideological battle against Communism. Bottling operations were started in Belgium, the Netherlands, and Luxemburg in 1947, but in France there was great opposition, first from the Communist party, which argued that they would become 'Coca-colonisés' and that the distribution network would double as a spy network, and second from the wine-growing, fruit-juice, and mineral-water interests. The French government, concerned by the trade deficit and the repatriation of profits, turned down requests by Coca-Cola to invest in France in 1948 and 1949, and banned the import of the ingredients from Casablanca. A bill was tabled by the deputy mayor of Montpellier on behalf of the winegrowers to empower the health ministry to investigate the content of drinks made with vegetable extracts in the name of public health. Its passage through the National Assembly in February 1950 provoked a storm of controversy. Farley visited the State Department and the French ambassador in Washington. The Americans put pressure on the French government. An article appeared in *Le Monde*, entitled 'To Die for Coca-Cola', mimicking the 'To Die for Danzig?' article of 1939. 'We have accepted chewing gum and Cecil B. De Mille, *Reader's Digest* and be-bop,' it read. 'It's over soft drinks that the conflict has erupted. Coca-Cola seems to be the Danzig of European culture. After Coca-Cola, enough.'[4] The French government was caught between the anger of French public opinion and the need to retain the favour of the American government. In the end the matter was resolved by the French courts, which ruled that the contents of Coca-Cola were neither fraudulent nor a health hazard. The French government retained its honour and the Americans obtained their market.

The fear of German resurgence

There was a view in France that having won the war in 1918 they had bungled the peace, and that this must not happen again after 1945. The risk of further German aggression must be eliminated once and for all. The French opposed the re-emergence of a centralized German state, and to this end tried to block the formation of centralized administrative services under the Interallied Control Commission. Though de Gaulle spoke of the end of the Thirty Years War, he really wished to go back 75 years and reduce Germany to the confederation of states that it had been before 1871. In any case, France should be given the territorial guarantees of security that Marshal Foch had demanded in 1919 but had been refused, in exchange for an American guarantee of the peace that was promptly rejected by the United States Senate. There were three clear demands on France's list: first, that the left bank of the Rhine be separated from the rest of Germany as a buffer state; second, that the coalfields of the Ruhr, the basis of Germany's military strength, be placed under the supervision of an international authority; third, that the coal-rich Saar, which had been in a French customs union between 1918 and 1935, and was now part of her zone of occupation, should once again be economically tied to France in a customs union.

Naturally the French did not have the military means to impose these demands, and could achieve them only by negotiation with the other powers. Bidault told Truman of the outlines of the French scheme at the San Francisco conference in May 1945, but no answer was either requested or given. He had to wait for the twenty-third session of the council of Allied foreign ministers in London, on 26 September 1945, before presenting his case, and received no support from Molotov, who might have been thought to favour dealing harshly with Germany. The main problem, as in 1919–20, was the United States. Despite some controversy within the administration, the dominant American view was that Germany should not be weak, divided, and 'pasturalized' but a strong state (even if only West Germany), economically strong in order to sustain the American economy, and militarily viable to act as a buffer against the Soviet Union. The United States, moreover, held the purse-strings of European recovery, so that while France's security needs dictated one course, her economic needs imposed another.

France tried to find a way out of the dilemma by forming a partnership with Moscow. The Soviets were keen to exact reparations from Germany, and to acquire an interest in an international authority supervising the Ruhr, neither of which was acceptable to Great Britain or the United States. France looked to support the Soviet Union's claims, if the Soviet Union in turn supported France's ambitions for economic union with the Saar. In New York, in December 1946, some sort of deal was negotiated by Molotov and Couve de Murville, director-general at the French foreign ministry. But the British and Americans seduced France by giving her what she wanted in the Saar, and at the Moscow conference of foreign ministers in March–April 1947 relations between Bidault and the Soviets finally broke down as the latter found they had no support on reparations and the Ruhr.

France's shift to the American camp was confirmed by the Marshall Plan, which provided aid for France but not (in the end) for the Soviet Union. The United States now pressed ahead with its programme of a constituent assembly in West Germany to frame a strong, centralized state, agreed at the London conference of June 1948, and French dreams of dismembering Germany dissolved. Some safeguards were projected to control a German resurgence, namely an International Authority of the Ruhr, to supervise the sharing out of coal supplies, on which Great Britain, France, West Germany, the Benelux countries, and the USA—but not the USSR—were represented, and a Military

Security Board to guarantee German demilitarization. Bidault was
squeezed between the realities of the international situation and public
opinion in France, which had been persuaded by the rhetoric of the
politicians that Germany was going to be dealt with in such a way as
to make any future revival impossible. He told the council of ministers
in May 1948, 'There is not the slightest shadow of a chance that we can
combine the benefits of Marshall Aid and refusal to accept a Germany
which in any case matches 50 per cent of our plans. If we want to go it
alone, we will lose everything.'[5] When he presented the London
agreements to the National Assembly, however, Bidault was crucified
by angry deputies who feared for French security and smelled betrayal.
Pierre Cot, formerly a Radical minister in the Popular Front government but now much closer to the Communists, said that 'the victims of
Nazi barbarism' must not be forgotten and warned of a 'renaissance of
the German peril'. Roland de Moustier, deputy for the Jura, described
Bidault's capitulation as 'the funeral oration of a policy' and regretted
that 'this policy, called the "policy of grandeur", has borne such bad
fruit'.[6] After this humiliation, Bidault lost his post as foreign minister.

Attempts at a European solution

The ambition of the Americans was not only to build a strong Federal
Republic of Germany but also to bring to an end the military
occupation. The Germans, meanwhile, chafed under the restrictions
and humiliations imposed upon them. In 1947 the Saar became part of
an economic union with France, but its German population voted
almost unanimously for political autonomy, acquiring their own government and assembly. The long-term aim of the Bonn government was
to recover complete control of the Saar, and it rejected a bid by the
French in January 1950 to obtain a fifty-year lease on the territory.
The Ruhr posed even more problems. The International Authority of the
Ruhr discriminated against the Germans and the Germans alone in
the matter of coal and steel production. They were keen, if coal and
steel production had to be supervised, that it should include industrial
areas outside the Ruhr.

The French realized that compromises would have to be negotiated
if they were not to be dictated to by the Americans. An imaginative
solution, that of a European Coal and Steel Community, was found by
two politicians, the new French foreign minister, Robert Schuman, and
the head of the economic planning agency, the Commissariat au Plan,
Jean Monnet. Schuman was admirably qualified to find a European

solution. Born in Luxemburg in 1886, the son of a Lorrainer who had
opted for German nationality after 1870, he studied in German
universities and served for a year in the German auxiliary services
during the First War. After Alsace-Lorraine was reunited with France
he was elected deputy for the Moselle between 1919 and 1940, cham-
pioning Catholic and conservative causes. Though he voted full powers
to Marshal Pétain and returned to German-occupied Lorraine, he was
placed under house arrest by the Germans, escaped to France where he
lay low, sat on the departmental Committee of Liberation in the
Moselle, joined the MRP, and became prime minister in 1947–8.
Schuman nevertheless had the farsightedness to judge that only a
European solution would prevent France and Germany tearing each
other apart for a third time over their borderlands, and the negotiating
skills to win agreement for the plan. Monnet, for his part, had the
confidence of the Americans and the necessary clout to sell the plan to
them. Their solution, revealed at a press conference in May 1950,
envisaged placing all the coal production of France and Germany—and
that of any other European country that wished to join them—under a
joint authority. The rhetoric of the declaration spoke of making future
war both unthinkable and materially impossible, and of the first step
towards a European federation. For the French it was also the
continuation of their Ruhr policy by other means, retaining access to
Ruhr coal and controlling German steel production, the mainspring of
her economic revival, even after the International Authority of the
Ruhr was no more.

While Great Britain had not been consulted on this initiative, the
Americans were well pleased. Indeed, the beginning of the Korean war
in June 1950 and the need to move troops from Europe to the Far East
caused them to put increasing pressure on Europe to see to its own
defence. Specifically, this meant the rearming of Germany and German
membership of NATO. The British and French foreign ministers, Bevin
and Schuman, were summoned to New York in September 1950, to be
told sharply by Secretary of State Dean Acheson that he wanted
Germans in uniform by the following autumn. For its part the Federal
Republic, faced by the deployment of heavily armed 'police' on the
frontier of the Democratic Republic of East Germany, was pressing for
the right to raise forces of its own. Bevin caved in before the American
demand. Schuman was vehemently opposed to the rearming of Germany, not least because at that moment the French army was committed in Indo-China. But he was isolated, and something more than a
blank refusal was required.

The answer, once again, was provided by Jean Monnet and put to Schuman and the prime minister, René Pleven. Its brilliance was to permit the rearming of Germans but not the rearmament of Germany. It also kept Germany out of NATO. It provided for the formation of a European army, under European political and military institutions. The National Assembly approved it in outline, without enthusiasm, by a majority of 343 to 225 with 31 abstentions in October 1950. Initially it angered the Americans, who disliked the complications of the political structure and saw it as a means to delay German rearmament, but Eisenhower was won over by Schuman in the summer of 1951. Talks on the European Defence Community, as it was called, opened in Paris in February 1951, without the participation of the British, and the final treaty was signed in Paris on 27 May 1952.

For the French, however, the trouble was only just beginning. The ratification of the EDC treaty was as controversial and divisive as that of the Maastricht treaty forty years later, if not more so. The elections of June 1951 moved the centre of gravity in the Assembly to the Right and made it very difficult to find a majority for ratification. The socialists, who were basically a European party, went into opposition, while the Gaullist Rassemblement du Peuple Français (Union of the French People, or RPF), violently hostile to a European army, became part of the right-wing governing coalition. As a price for this favour it demanded that the EDC treaty be put on ice and that Robert Schuman be dismissed from the Quai d'Orsay. The ministry of the Radical Mendès France in June 1954 shuffled the pack, but to no better effect. The RPF resigned from government, while the MRP, the most European of the parties, was opposed to him. Mendès France took the view that the Assembly could no longer avoid debating the treaty, which had been ratified by the other European partners concerned, but protected his own government by refusing to make ratification an issue of confidence and, indeed, refusing to speak in the debate.

The debate was heated in both parliament and the country. Supporters of the EDC argued rationally that German rearmament would come sooner or later, and that the EDC offered an institutional way to control it; it would also keep Germany out of NATO by a European solution linked to NATO. Unfortunately, the debate was in no sense rational, and opponents of the EDC were able to prey on painful memories of the German occupation and of Nazi atrocities, revived once again in 1953 in the Oradour trial. They argued that the EDC would not control German rearmament but actually resurrect the Wehrmacht, and that the Wehrmacht, as in the Bismarck era, would forge a united German Reich. Conversely, they argued that the EDC would cut the French army in two, leaving its colonial army on one side, and destroy not only the autonomy of the French army but also its soul, lost in an artificial, stateless body. All narrowly nationalist and anti-federalist opinion converged to fight the EDC treaty.

In the country, opinion was divided equally and immovably into three camps. Six polls, held between May 1953 and January 1955, showed a third in favour of the EDC, a third against it, and a third undecided. That the feelings aroused by the Oradour trial ran strong was demonstrated by the fact that the only region overwhelmingly in favour of the European army was Alsace, while the centre-west, which included the Limousin, was second only to Paris in its opposition. The parties were similarly divided. Only the MRP was decisively in favour of the treaty, with 60 per cent of its supporters persuaded. It was approved by 50 per cent of supporters of the Right and 44 per cent of Radical supporters. Among Communist supporters, a full 78 per cent were opposed to the treaty, while socialist and Gaullist supporters reflected the national split into three more or less equal groups. When the treaty was debated in the Assembly at the end of August 1954, the 82-year-old Radical Édouard Herriot played on the harp-strings of history. France, he said, could not accept a supranational army controlled by robots. 'The army is the soul of the fatherland . . . it is because the feelings developed by the French Revolution had such depth that they were able to give the men who fought on the Marne the courage to die in conditions that we must not forget.'[7] All the Communists, most of the Gaullists and, decisively, half the Socialists voted against the EDC treaty. It was rejected by 319 votes to 264, the result acclaimed by shouts of 'Down with the Wehrmacht!' and singing of the 'Marseillaise'.

As the supporters of the EDC had predicted, Germany joined NATO and began to rearm anyway. But the scuppering of the EDC was a major blow to progress towards European federation. Something had to be salvaged from the wreckage, but there could be no question of any political or military union. Jean Monnet wanted to take advantage of the United States' new willingness to provide technology for the peaceful use of nuclear power to establish an atomic community, Euratom, both to develop new forms of energy and to control the nuclear projects of the Federal Republic. The Germans, on the other hand, were keen to establish a common European market to sustain their economic miracle, much to the annoyance of the majority of French people, who wanted to protect their somewhat traditional economy and society against the cold winds of competition behind high tariff walls. There was also, however, an alternative opinion among technocrats in France that she must imitate the American model of high productivity and a high standard of living, and this required the economies of scale that only a wide European market could provide.

The foreign ministers of the member states of the Coal and Steel Community met at Messina in June 1955 (Great Britain declined to attend), and negotiations were continued by an intergovernmental committee in Brussels. The report produced under the name of the Belgian foreign minister and former premier Paul-Henri Spaak combined the ideas of an atomic community and a common market, and the foreign ministers signed the resulting treaty of Rome in March 1957. The French had obtained a whole string of concessions to induce them to sign, including a high external tariff, exchange controls, the association of her colonies, and a common agricultural policy to subsidize farmers. The election of January 1956, which inflicted defeat on the anti-European Gaullists and brought in a socialist-led government with a pro-European foreign minister, Christian Pineau, ensured an easier ride through the National Assembly than that experienced by the EDC. Despite the opposition of Communists, Gaullists, and the Poujadist Right, and a hostile speech by Mendès France warning of the dangers of German industrial hegemony and immigrant Italian labour, the treaty was ratified by 342 votes to 239 in July 1957. And despite a decade of anti-European rhetoric by General de Gaulle, he was wise enough to accept the European Economic Community when it came into force on 1 January 1959.

In search of a lost empire

'Without the Empire, France would not be a liberated country. Thanks to its Empire, France is a conquering power.'[8] Thus Gaston Monnerville, an assimilated black from French Guiana and future president of the Conseil de la République and Senate, addressed the provisional Consultative Assembly on 25 May 1945. The Empire, which had remained out of German hands in 1940, and control of which the Free French had gradually wrested from Vichy, served as a springboard for the liberation of metropolitan France, both strategically and in terms of the colonial troops made available. Nearly 300,000 North African Arabs, for example, fought in the ranks of the Free French. Subsequently, possession of the Empire served as the basis of the French claim for great-power status, *vis-à-vis* Great Britain, the rival colonial power, and the United States, the dominant superpower. It is not surprising, therefore, that immediately the war against Germany had been won in Europe, France was keen to re-establish her imperial power in Africa, the Levant, and Indo-China.

There was, unfortunately, a contradiction between France's great-power ambitions and the universal mission she saw as hers to liberate and civilize oppressed and benighted peoples. Fighting against the tyranny of the Axis powers to liberate herself, she could scarcely deny liberty to others in her charge, especially as they came to expect liberation from the new France that emerged from the Resistance. In March 1943, for example, Algerian nationalists under Ferhat Abbas issued a *Manifesto of the Algerian People* demanding an autonomous Algerian state. De Gaulle organised a conference of colonial governors (not nationalist leaders) at Brazzaville in January 1944 to sketch out the framework of the Union that would supersede the Empire, and which gave the impression of liberation at the hands of France. But liberation and civilization tended to pull against each other, the French arguing that more liberty was not due until greater civilization had been achieved. The Brazzaville conference thus concluded that 'the goals of the work of civilization undertaken by France in the colonies exclude all idea of autonomy, all possibility of development outside the French bloc of the Empire; the possible constitution of self-government in the colonies is to be dismissed.'[9] The sort of liberty the French were prepared to consider in their colonies, as exemplified by that given to the Algerians in March 1944, included wider civil liberties for Arabs, who were not considered fully French citizens, and a larger electorate in the Arab college, which was to remain separate from the European college. A small minority of Arabs were allowed to vote in the European college, in respect of their proven assimilation, for one aim of the civilizing mission was to undermine nationalist claims by creating an assimilated Francophile élite which would see liberty as deriving from participation in French democracy and culture. The French could not understand that the liberty they offered was not entirely sufficient for colonial peoples. Where disturbances broke out in overseas territories at the Liberation in support of demands for independence, the French authorities blamed them either on Nazi agents or on British intriguers. Whatever promises may have been made to colonial peoples during the war, to win them over to the Free French and to complete the victory against Germany, there was never any intention of surrendering the material advantages that derived from colonial power.

An article of the constitution of the Fourth Republic, which restated that of the constitution of 1791, promised that France would never undertake a war of conquest or use force against the liberty of any people. The methods used by the French to reimpose colonial power suggest either that colonial wars were not envisaged by this profession of faith or that the inhabitants of French colonies were not considered to be people. When the French blocked the demands of moderate nationalists in Algeria, the initiative passed to the radicals under Messali Hadj. The French deported Messali on 25 April 1945, which provoked demonstrations on 8 May 1945. At Sétif riots broke out, during which over 100 French settlers were massacred. The French replied with brutal repression, killing between 1,500 and 8,000 Algerians according to French sources, 45,000 according to Algerian ones. At the end of the same month the French bombarded Damascus, killing hundreds. The bombardment of Haiphong on 23 November 1946 killed about 6,000. The repression of the insurrection in Madagascar, which began in March 1947, killed 11,200 officially, and possibly between 100,000 and 200,000. As in the French revolutionary wars, the French brought 'liberty' and 'fraternity' at the point of the bayonet and from the barrel of the gun.

Folies de grandeur: Syria-Lebanon and Indo-China

The French *folies de grandeur* might have been excusable if they had not been so obviously *folies*, backed up by neither military strength nor political will. In June 1941 the Free French under General Catroux made an attempt to retake Syria and Lebanon from the Vichy regime of General Dentz. But most of the fighting was done by the British, Australians, and Indians, and Catroux was not even allowed to sign the act of surrender. In September 1941 Catroux published a manifesto promising Syrian independence, but continued to exercise power as high commissioner and treated Syria and Lebanon like client states. Immediately they had the means, on 6 May, the French sent a cruiser to Beirut and demanded military and economic concessions from Syrian and Lebanese ministers in Damascus. French terms were rejected, a general strike was proclaimed, and French encampments were attacked. The French replied by shelling Damascus on 29, 30, and 31 May, until the British arrived with tanks and confined French forces to barracks. 'We are not, I admit, in a position to open hostilities against you,' de Gaulle told the British ambassador Duff Cooper, 'but you have insulted France and betrayed the West.'[10] The French and British agreed to withdraw their forces jointly, and left in 1946.

In Vietnam the Vichy regime under Admiral Decoux was finally toppled by the Japanese in March 1945, and replaced by that of the former Emperor of Vietnam, Bao Dai. But Bao Dai was no match for the Communist Vietminh of Ho Chi Minh, who swept to power as the Japanese faced defeat and declared an independent Vietnamese Republic in September 1945. At Potsdam, where the French were not present, the Allies agreed that the British would disarm and repatriate Japanese forces south of the 16th parallel, while the Chinese nationalists did the same to the north. There was no French military presence in Vietnam until October, when General Leclerc arrived in Saigon. Admiral Thierry d'Argenlieu, a former monk and admirer of Cardinal Richelieu, followed as High Commissioner at the end of the month. Their mission was to recover not only Vietnam but also Cambodia and Laos, and to group them into an Indo-Chinese federation within the French Union. Unfortunately the French started out with only 4,500 troops, rising to 30,000 in January 1946. Unable to enforce their will they were obliged to negotiate with the Vietnamese and Chinese, and in March 1946 agreed to recognize the independence of the Vietnamese Republic within the Indo-Chinese federation and French Union, accept a referendum to unify the constituent provinces of Tonkin, Annam, and Cochin-China, and to withdraw French forces after five years. Because the French could not win by force, they tried fraud. In June 1946, the day after Ho Chi Minh had left for talks in France, d'Argenlieu authorized the proclamation of a separatist Cochin-Chinese republic at Saigon, and bombarded Haiphong from the sea.

In the ensuing war there were never more than 100,000 French in Vietnam, of whom a third were civilians. By 1947 they controlled the main cities and roads, but the countryside answered to the Vietminh. To recover some legitimacy they installed Bao Dai as the ruler of a phantom 'independent' Vietnam within the French Union, but he was no more than a puppet and in any case could not be induced to give up his playboy life-style in Cannes. After 1950, what started as a colonial war became complicated by the Cold War, as Communist China and the Soviet Union recognized the Democratic Republic of Vietnam and began to support its war effort. In France the Communists switched from the line they had followed from 1945, preaching the orthodoxy of France's mission of liberation, civilization, and greatness, to a vehement anti-colonial campaign. Jeannette Vermeersch, partner of the PCF leader Maurice Thorez, railed for two hours in the

Assembly against French action in Vietnam, while CGT dockers tried to stop the shipment of military hardware.

After a military disaster at the fortress of Cao Bang in October 1950, French strategy reached a turning-point. On the one hand were the partisans of an increased war effort, sending the veteran Second World War general de Lattre de Tassigny out to Vietnam both as high commissioner and commander-in-chief, recruiting Vietnamese into a national army of 150,000, and putting pressure on the United States, themselves involved in Korea, to provide aid. On the other hand there were partisans of negotiation like Pierre Mendès France, who argued that the French would have to commit three times as many troops as they had to have a chance of military success, but that the war was destroying France's prospects of economic modernization and competitiveness in world markets. Mendès France founded the weekly *Express* in May 1953 to put across his radical views. Asked by the president of the Republic to form a government in June 1953, he was rejected by the Assembly. It was not until after the military disaster of Dien Bien Phu, a vital strategic point taken by paratroops in November 1953 but recaptured by General Giap on 7 May 1954, that Mendès France was able to form a government. To obtain the approval of the Assembly he requested a mandate of one month to negotiate an armistice, offering his resignation if he failed. Flying to Geneva, he used the pressure of the one-month deadline to persuade the Soviet foreign minister Molotov into accepting an armistice line on the 18th rather than the 16th parallel, and returned with the deal just in time.

Algeria: the war without a name

'There was a prime minister whose name was Mendès, and he also had another name, but he was too small for such a big name. It was he who made France lose enormous territories.'[11] This tirade against Mendès France in the Assembly in December 1954 was typical of the anti-Semitic abuse of which he was so often the target. He was variously attacked as a capitulator in the hands of the Rothschild bank and an Oriental carpet-dealer selling off France's empire on the cheap. True, he took France out of Vietnam and tried to reconcile the conflicting demands of nationalists, French settlers, and the French state in the North African protectorates of Tunisia and Morocco. He visited Carthage in July 1954 to promise Tunisian internal autonomy and allowed the return to Morocco of the sultan, who had recently been ejected by diehard French settlers. It is likely that Algerian nationalists hoped to gain concessions from Mendès France when they formed a National Liberation Front (FLN) and National Liberation Army in October 1954, and on 1 November launched an insurrection in the mountainous south of the country in pursuit of a sovereign Algerian state and the abolition of all distinctions based on religion or race. Here, however, Mendès France drew the line. Algeria, a French colony since 1830, was divided into three departments and considered part of metropolitan France, under the jurisdiction of the ministry of the interior. Mendès France was determined to 'maintain the unity and indivisibility of the Republic, of which Algeria is a part', and in January 1955 appointed the tough former Resistance leader, Jacques Soustelle, governor-general of Algeria.[12] Soustelle, affirming a policy of 'integration', argued that 'It is precisely *because* we have lost Indo-China, Tunisia and Morocco that we must not, at any price, in any way and under any pretext, lose Algeria.'[13] Even so, Mendès France was regarded as someone who might lose Algeria, and his government was overturned in February 1955.

The Algerian war was one of the most tragic episodes of twentieth-century French history. The tragedy was that the French considered that they were doing the right thing in Algeria, and had no understanding that what they were up against was a post-colonial war of national liberation. Though the fighting continued for eight years, as far as the government was concerned there was no war in Algeria, only internal problems of public order, referred to as 'events'. Censorship was tight under emergency legislation, and newspapers which so much as spoke of war were liable to be seized. Songs and films dealing critically with the war were also censored: thus Boris Vian's 'Le Déserteur' of 1954 was banned, while Jean-Luc Godard's film *Le Petit Soldat*, made in 1959, was not shown until after the war, in 1963.

The problem was made more difficult by the presence in Algeria of a large and vociferous French settler population, nearly a million strong in 1954, called *pieds-noirs* because their polished black shoes distinguished them from native Algerians who tended to go barefoot. Like the Protestant Unionists in Northern Ireland, they regarded any concessions made to the other side, in this case the Arab Algerians, as a threat to their position and to the union with France. The elections of January 1956 were won by a Republican Front led by the socialist Guy Mollet, which looked to end hostilities by political and economic reforms that would rally the silent majority of Arabs to French institutions and isolate the nationalists. But as Mollet was laying a

wreath on the war memorial of Algiers on 6 February 1956 he was
pelted with tomatoes by angry *pieds-noirs*, and promptly reversed his
policy. He appointed a hardliner, the socialist deputy Robert Lacoste,
minister-resident of Algiers, and equipped him with almost dictatorial
means under a special-powers law passed by the National Assembly on
12 March 1956 for his 'policy of pacification'. Lacoste used his powers
with devastating effect, for example hijacking a plane in which FLN
leaders were travelling. But as the *pied-noir* Albert Camus told students
in Stockholm in December 1957, as he collected the Nobel Prize for
literature, 'I believe in justice, but I would defend my mother before
justice.'[14]

In theory, the presence of a government of the Left should have
induced a more understanding, humanitarian approach to the problem.
If anything, however, parties and intellectuals of the Left were even
more fervent believers in the universal liberating and civilizing mission
than those of the Right. It was the responsibility of the French Republic
to provide individual liberty, democracy which progressively brought
Muslims into the fold of citizenship, secularization instead of religious
fanaticism, civilization instead of medieval backwardness. These were
the tenets of Mollet's socialists, of Radicals like Mendès France who
served briefly in his government, and even of the Communists, who for
the first time since 1947 supported the governing majority and sealed
the partnership by voting the law on special powers. 'It is abominable
to hear said by some twisted minds that there is some sort of
comparison or analogy between the present Algerian rebellion and the
former French Resistance,' wrote one socialist in 1957. 'It is beyond all
doubt that the French army is the logical continuation of the action of
the French Resistance as a whole.' Similarly, the heirs of those who had
put justice for the individual before *raison d'état* in the Dreyfus Affair
of the turn of the century were now on the side of Algeria for the
French.[15] Albert Bayet, president of the Ligue de l'Enseignement, and
the anthropologist Paul Rivet, one of the founders of the Vigilance
Committee of Anti-Fascist Intellectuals in 1934, along with Jacques
Soustelle, were among the signatories of a manifesto which denounced
'the instruments of theocratic, fanatical and racist fanaticism' and
asked, 'who, if not the *patrie* of the rights of man, can clear a human
way to the future' for the populations of Algeria?[16] Given the consensus
of the major parties and leading intellectuals behind the war, and their
decisive appropriation of the myths of the Dreyfus Affair and the
Resistance in support of their cause, opposition had to develop on
the margins, in reviews such as Claude Bourdet's *France-Observateur*,

in the publications of the Catholic Left such as *Esprit* and *Témoignage chrétien*, and in the student movement.

At the grass roots, opposition to fighting the war also came from the conscripts themselves. Unlike the war in Indo-China, those doing national service were called upon as well as the professional army. Particularly unpopular was the recall, under the special-powers act, of 70,000 young men who had already done their military service and returned to civilian life. Riots broke out in April, May, and June 1956 at railway stations across the country as demonstrators tried to prevent the trains carrying the *rappelés* from leaving. But there was little support for these protests from the parties and unions, even the Communists, and President Coty made a speech at Verdun in June 1956 to warn that in Algeria the fatherland was in danger, and that undermining the morale and discipline of those sent by the Republic to combat terrorism was quite unacceptable. The heat of patriotism was turned up even higher during the Suez crisis. Nasser of Egypt was held responsible by the French government for training and supplying the FLN, and when he nationalized the Suez Canal Company in July 1956, the French considered that the FLN could not be defeated until Nasser had gone. Again, historical analogies were found: Nasser was Hitler, and the seizure of the Canal the German reoccupation of the Rhineland in 1936, except that this time the French would not be caught out. Despite American opposition, the French pressed Great Britain and Israel to join in military action, which was voted by 368 deputies 'with a light heart', in the words of the opposition press, evoking the gung-ho attitude of the Franco-Prussian war of 1870, and supported by a majority of French public opinion.

Just as patriotism was whipped up to conceal what was really the imposition of internal order, so the army was portrayed as an instrument of the civilizing mission when it was in fact engaged in a campaign of systematic repression and torture. Pictures were published to highlight the role of the Special Administrative Sections, which helped with education and agriculture in 'friendly' Algerian villages. The main function of the army in the countryside, however, was to seal off and comb villages in pursuit of FLN supporters. Moreover, following a number of terrorist attacks in Algiers, Robert Lacoste handed over his emergency powers in January 1957 to General Massu, commander of the 10th Paratroop Division, who then waged a campaign of terror in the city to flush out the rebels. Since the notion that the FLN were an isolated fanatical minority had broken down, as it was realized that they had the broad support of the population, so torture was stepped up by the paratroopers in order to track down their prey.

The scandal that erupted over torture after 1957 exposed the cult of the French liberating and civilizing mission for the sham that it was and precipitated a painful reconsideration of French national identity. The myth of the French Resistance had reinforced the view that the French were always on the side of liberty and justice against oppression and injustice, but now it was demonstrated, little more than a decade after the Occupation, that in Algeria the French were using the same tactics as the Gestapo. 'Your Gestapo in Algeria', screamed Claude Bourdet in *France-Observateur* as early as January 1955, when the perpetrators were the police rather than the army. In March 1957 Paul Teitgen, secretary-general of the prefecture of Algiers, complained to Massu that he had seen traces of torture reminiscent of those he had suffered fourteen years previously in the cellars of the Gestapo in Nancy, and offered his resignation, though this was turned down. The following month Guy Mollet vehemently rejected any comparisons of the French army with the Gestapo as 'scandalous. Hitler gave instructions advocating these barbaric methods, whereas Lacoste and myself have always given orders in an absolutely contrary sense.'[17] Whether or not Mollet did give the order to torture, the fact is that in Algeria in 1957 Massu's paratroopers made the law. The cover-up was finally blown by a series of *causes célèbres*, as French sympathizers as well as Algerians were subjected to torture. In June 1957 Maurice Audin, a lecturer in the science faculty of Algiers, was arrested by paratroopers and disappeared. An Audin Committee was set up by the historian Pierre Vidal-Naquet to discover the truth. At the same time the Communist Henri Alleg, editor of *Alger républicain*, which was banned in 1955, was arrested by paratroopers and tortured. He revealed his experiences in *La Question*, published in February 1958 with a preface by Sartre and which sold 60,000 copies before it was clumsily banned six weeks later. Germaine Tillion, an anthropologist who had been deported to Ravensbrück as a member of the Musée de l'Homme Resistance cell, and then served in the office of Governor-General Soustelle with a brief to help develop Algeria, subsequently recalled how she had intervened without success to try to stop the execution of ten of her comrades in 1941–2 and in July 1957 tried—again without success—to stop the execution of FLN militants. It was now clear to many intellectuals that the victim had become the executioner, and that the French mission in Algeria had lost every shred of legitimacy.

This loss of nerve, which soon affected French politicians too, drove the *pieds-noirs* to desperate measures. Establishing links with extreme right-wing organisations and disgruntled army officers, they organized a mass strike in Algiers on 13 May 1958, invaded the governor-general's building and set up a committee of public safety under General Massu. They won over General Salan, who was made governor by Paris in the hope that he would maintain order, and called upon General de Gaulle to set up a government of public safety to save French Algeria. De Gaulle kept his distance from the organizers of the coup, orchestrated his return to power by legal and constitutional means, and received a popular mandate as president of a new Republic. But he would not have been able to return to power without the crisis precipitated by the organizers of the coup in Algiers, and became the object of their wrath when, within a short space of time, he began to feel his way towards self-determination for Algeria.

In a broadcast of 16 September 1959 de Gaulle announced that the only course open to a 'great nation' like France was to offer self-determination to the Algerian people. Within four years of a ceasefire they would be given the choice between 'secession' (independence), 'Francization' (integration), and 'the government of Algeria by Algerians, supported by French aid and in close union with France' (association). It was clear that de Gaulle, anxious to strike before the United Nations voted on independence for Algeria, favoured the third option, and equally clear that ultras in the army and the *pieds-noirs* would not budge from the second. When Massu was recalled to Paris in January 1960 for attacking self-determination, *pied-noir* extremists launched a general strike in Algiers in what became known as 'the week of the barricades'. De Gaulle appeared on television in military uniform to recall the rebels to order. He pressed on with his policy, calling upon the 'liberating genius of France' which in 1960 was in the process of emancipating her sub-Saharan colonies, and obtained 75 per cent support for the principle of self-determination in a referendum of January 1961. Before talks could open with the Algerian provisional government, opposing generals led by Salan launched a putsch in Algiers in April 1961 which was quashed only with difficulty. The struggle for French Algeria was then pursued by the Organisation de l'Armée Secrète (OAS), which linked dissident soldiers and political extremists and waged a campaign of terror both in Algeria and on the mainland, including attempts to assassinate de Gaulle himself.

Despite their counter-revolutionary stance and terrorist methods, these extremists had no doubt that they had right on their side. On trial, they invariably proclaimed that they were defending French honour against the treachery of de Gaulle and the politicians. Court-martialled as head of the OAS in 1962, Salan argued that he was fighting to preserve the empire of Galliéni and Lyautey and that veterans of the war in Indo-China were not going to lose again in Algeria. Georges Bidault, who had headed one National Resistance Council in 1943 to fight the German Occupation, set up another in 1962 to prevent the abandonment of Algeria, and regretted that he was having to start a new career in the Resistance at the age of 60. Others, however, contested that the legacy of the Resistance should be appropriated by the OAS. In May 1959 a group of draft-evaders and deserters set up an organization called Jeune Résistance. They refused to fight a war that perverted French values of liberation and civilization. They wanted not only to stop the war but to overthrow de Gaulle, whom they saw as a tool of militarism and fascism, in order to found a new regime based on peace and socialism. Others went even further, arguing that the appropriate revolutionary position was to help the Algerians with their war. Led by a philosophy teacher, Francis Jeanson, they set up a network to handle FLN funds in France. Arrested and put on trial in September 1960, Jeanson and his so-called bag-carriers for the FLN were supported by 121 intellectuals headed by Jean-Paul Sartre. Their controversial manifesto, which concluded that 'the cause of the Algerian people, which contributes decisively to undermining the colonial system, is the cause of all free people', abandoned the flawed liberationist myth for a new Third World position.[18] Sartre underlined his position by writing a preface to Frantz Fanon's *Wretched of the Earth* (1961), an indictment of the physical and mental violence perpetrated by the French colonial system, particularly in Algeria.

As the campaign against the Algerian war gained momentum, so the violence of the French state spilled back onto the mainland. On 17 October 1961, tens of thousands of Algerians demonstrated in central Paris in favour of peace, negotiation, and an Algerian Algeria. The police charged, killing two, wounding 69, and making 11,538 arrests. At the pont de Neuilly police trapped and beat Algerians and threw them into the Seine; 60 bodies were recovered over the next month. Sartre denounced a 'police pogrom' undertaken on the orders of the Paris prefect of police, Maurice Papon, who would subsequently face charges of crimes against humanity for his deportation of Jews during the Occupation.[19] At the time much was done to cover this up, and more publicity was received by a demonstration organized by the trade unions and parties of the Left on 8 February 1962 against OAS violence and for peace in Algeria. As the demonstration broke up the police charged at the Métro station Charonne, killing eight—all members of the CGT and seven of them also Communist militants—and injuring hundreds. A general strike was called for the day of the funeral, 13 February, which was attended by half a million people.

Soon, however, the violence of the state turned in the opposite direction, against partisans of French Algeria. A ceasefire agreement was signed at Évian in March 1962 and a referendum on 'Algerian independence in co-operation with France' was approved by 90 per cent of voters in metropolitan France on 8 April 1962, by 99 per cent of voters in Algeria on 1 July. Before this, on 26 March, the OAS proclaimed a general strike in Algiers. Provoked by the OAS, the French army fired on French Algerian demonstrators in the rue d'Isly, killing 66 and leaving 200 wounded. After independence about 150,000 *harkis*, those Algerians who had remained loyal to and fought alongside the French, were killed by the FLN, as were up to 10,000 *pieds-noirs*. The *pieds-noirs* were repatriated from Algeria, loath to leave what they considered their homeland, while surviving *harkis* were crowded into camps in the south of France.

After independence, an Algerian constituent assembly proclaimed a Democratic and Popular Algerian Republic. The FLN was established as a single party of government and its leader, Ben Bella, elected president in 1963, undertook a programme of agrarian reform and forced industrialization inspired by Nasser's Egypt. The army of General Boumedienne was, however, the real power behind the throne and in 1965, after a disastrous war against Morocco, it removed Ben Bella and made Boumedienne president. The French found it easy to do business with a regime that ensured order and modernization and retained French as an official language alongside Arabic. However, Arab nationalism in Algeria did not escape the rise of Islamic fundamentalism, and in June 1990 the Islamic Salvation Front (FIS), formed the previous year, triumphed in local and departmental elections. The Algerian government quashed the elections and dissolved the FIS in 1992, but its armed wing, the Islamic Salvation Army (AIS), mounted a campaign of terror both against the government and against French nationals whose government was seen to be supporting the military-backed regime in Algiers. Five French nationals were assassinated by the AIS in August 1994, and at Christmas 1994 the campaign spread to the French mainland when an Air France airbus was hijacked in Algiers by the Islamic Armed Group (GIA). The intention of the hijackers was to crash-land the plane on Paris, but they were over-

whelmed in Marseille by an anti-terrorist unit. The AIS promptly
declared war on France on behalf of the Algerian nation. The Algerian
war now crossed the Mediterranean once again, but this time the
terrorist groups in France were not the OAS but Islamic fundamental-
ists.

Algeria: the impossible memory

The Algerian war was impossible to remember in a way that integrated
happily with French history. For a long time, in the first place, it was
not accepted as a war but only as 'events' or troubles disturbing internal
order and requiring 'pacification'. The National Federation of Veterans
of the Algerian War (FNACA), set up in 1958, was unable to obtain a
veteran's card for its members until 1974. Second, the Algerian war was
lost, and represented the end of the French Empire. Not only that, a
poll for *L'Express* in October 1979 showed that 58 per cent of French
people regretted that the war had been fought, since the outcome
had been inevitable. Even so, third, Algeria was not France's Vietnam
because it was a civil war. The Algerians against whom war was
waged were regarded as Frenchmen, even if they were bad ones, and
terrorism had been brought to the streets and metro stations of Paris
by the OAS.

Attempts were made to reforge national unity. A first amnesty law
was passed as early as December 1964, and 173 OAS pardoned. The
need to win over conservatives for the elections that followed the events
of May 1968 hastened the pardon of all members of the OAS: Georges
Bidault returned from exile and Salan came out of prison. In 1982 the
socialist government allowed those who had been punished to take up
civil or military posts from which, until then, they had been banned.
But unity had to be forged around a collective memory, and there was
argument about what that memory should be. In 1971 the FNACA
began a campaign to secure commemoration of the Algerian war on 17
March, date of the Évian accords, as the end of the last great war of
the twentieth century. However, they were opposed by the main
veterans' organization, which argued that it was impossible to com-
memorate a sell-out and defeat, and by the *pieds-noirs*, who founded a
Committee for the Respect of the Memory of Those who Died for
French Algeria in 1981, chaired between 1986 and 1988 by a National
Front deputy. The clash came to a head after the annual FNACA
service at the Invalides on 19 March 1988, when eggs were rained on
participants by demonstrators whose placards recalled the 10,000

pieds-noirs and 150,000 pro-French Arabs who died for French Algeria. Lastly, the Algerian war challenged and undermined the values that for so long had defined French national identity: those of greatness and her liberating and civilizing mission. The French had failed to see that liberation might also mean the rejection of French definitions of liberty and civilization in favour of others. Moreover, to retain those values they had resorted to methods of barbarism which allowed them, the people of 1789, to be attacked for being no better than Nazis.

2

Crisis in the State

A revolutionary situation

In 1944 there was a revolutionary situation in France. It was the likeliest opportunity for a Communist seizure of power that the country had seen or would ever see. The German army of occupation was in retreat. The Vichy authorities that had been propped up by it were discredited and in collapse. The French Committee of National Liberation, which became the provisional government only just before the Normandy landings, was based in Algiers with a provisional Consultative Assembly, and had yet to assert its control in metropolitan France. The key presence on the ground were the combined forces of the Resistance, the Forces Françaises de l'Intérieur (French Forces of the Interior, or FFIs), backed up by *milices patriotiques* of citizens mobilized for national insurrection. The dominant element in the FFIs were the Communist Francs-Tireurs et Partisans (Irregulars and Partisans, or F.-T.P), and the FFIs as a whole were commanded by the Military Committee of the National Resistance Council (COMAC), two of whose three members were Communists. As towns and villages were liberated by these forces so new revolutionary authorities were set up, the Comités de Libération (Liberation Committees, or CDL). Established at commune and departmental level, it has been calculated that Communists formed 26 per cent of their membership in the former occupied zone, 35 per cent in the former unoccupied zone, the rest of the membership being made up of socialists, Christian Democrats, and Gaullists. Early in September 1944 the representatives of six departmental CDLs in south-eastern France met at the château of Vizille, outside Grenoble, where the French Revolution is said to have started in 1788, to federate together so as to be in a position to dictate terms to officials of the provisional government. These revolutionary forces played a key role in the liberation of Paris, and the National Resistance Council took up post in the *hôtel de ville*, the temple of all French revolutions, in order to receive General de Gaulle on 25 August.

Why, then, was there not a Communist revolution in France in 1944? The first part of the answer is strategic. France was ultimately liberated more by the Allies than by the forces of the internal Resistance, and the Americans, who faced having to cede eastern Europe to Stalin, were not going to permit Communist seizures of power in western Europe. On 27 August, indeed, Eisenhower was asked by de Gaulle to supply two American divisions to establish the authority of the latter in liberated Paris. Second, de Gaulle himself made the 'restoration of the state', that is, that of the provisional government, the first priority in liberated France. From the spring of 1944 he had sent prefects and *commissaires de la République*, with authority over a region of several departments, to wrest control from and ultimately dissolve the revolutionary authorities. In the short term, however, co-operation was more usually the order of the day, and the CDLs did not finally disappear until the election of new municipal councils in April and May 1945 and of new *conseils généraux* or departmental councils that autumn. A decree of 27 September 1944 integrated the FFIs into the regular army, composed of the Free French and others recruited after the Liberation, although most of the FFIs simply disbanded and went home. The real trial of strength came with the decree of 28 October 1944 ordering the disbandment of the *milices patriotiques*. If the Communists were going to attempt a show of force, it was these *milices* that would be their instrument. A rally of 100,000 of them was held in Paris on 11 November 1944 as an act of defiance, but at the end of that month the Communist party leader, Maurice Thorez, just back from Moscow where he had spent the war, told a rally at the Vélodrome d'Hiver that the Communist priority was not revolution, but war and victory against Germany. In January 1945 he told the Communist Central Committee that the slogan was 'a single army, a single police, a single administration', and the order was given for the *milices patriotiques* to dissolve.[1]

The third reason for the absence of revolution was thus the policy of the Communist party. Initially, it maintained a 'dual power' of the revolutionary authorities exerting pressure on the provisional government, as in Russia in 1917. A strong minority in the PCF was in favour of a revolutionary seizure of power, but Thorez was able to contain the pressure, not least by providing power in the state and legitimacy in the nation, without the need for revolution. As a leading force in the Resistance, they had a strong claim to office in the provisional government itself. Fernand Grenier and then, after September 1944, Charles Tillon, chief of the FTP, became de Gaulle's commissaire for air, while

François Billoux was commissaire for public health. Equally, just as the Nazi–Soviet pact of 1939 had meant ruin for French Communists, so entering the Resistance in 1941 when Hitler invaded the Soviet Union allowed them to reconcile their revolutionary and patriotic instincts and insist, after the war, that with their '75,000 martyrs' for France, they were the most patriotic party. More cynically, it suited Stalin after Yalta that France should regain its great-power status as soon as possible in order to limit the hegemony of the Anglo-Saxon powers in the West, and he had duly received de Gaulle in Moscow in December 1944 to conclude a pact. 'France, once more a great nation, must have a policy corresponding to its rank, and must take an increasingly active part in the war.' These words, addressed the same month to the Consultative Assembly, were spoken not by de Gaulle but by the leading Communist Jacques Duclos.[2]

The risk of dictatorship

When de Gaulle spoke of the restoration of the state, he had in mind not only the restoration of the centralized bureaucracy but also the rejection of the polity of the Third Republic, with its divisive parties, all-powerful parliament, and weak executive, which had led to disaster in 1940, and its replacement by a strong presidential regime based on popular appeal. Michel Debré, a former *commissaire de la République* now working in de Gaulle's private office, argued in a pseudonymous pamphlet in 1945 that 'the only chance for French democracy is, if the term may be used, a republican monarch'.[3] Though many French people at the Liberation would have agreed that the Third Republic had been a disaster, few followed de Gaulle's argument that what was required was a more authoritarian regime. An ordinance issued by the provisional government in Algiers on 21 April 1944 had promised a Constituent Assembly elected both by men and, for the first time, by women, and neither were prepared to pass up this opportunity to participate in building a new Republic. Many of the old politicians had been cleared away by the purges, opening the way for a new generation of politicians—men and women—who had won their political spurs in the Resistance and sought a public role in the new order. Lastly, though the Resistance was an extraordinary movement, and recast political identities and alignments in exciting new ways, it proved impossible to translate it into a political party, not least because both old political parties that had been extinguished under Vichy and new ones, founded at the Liberation, were keen to organize and compete for power.

Ironically, de Gaulle had favoured the restoration of political parties because he needed both to offset the influence of the Communist party and to demonstrate to the Allies the breadth of his support. Their representatives sat on the National Resistance Council, in the provisional Consultative Assembly at Algiers, and in the Consultative Assembly proper, which opened in Paris on 11 November 1944. From the autumn of 1944 to the summer of 1945, parties held conferences to constitute or reconstitute themselves. Initially the situation was very fluid. The Communist party and Socialists of the Section Française de l'Internationale Ouvrière (French Section of the Socialist International, or SFIO) met to see whether there was a chance of reconstituting the united proletarian party that had broken apart at the Congress of Tours in 1920. But Léon Blum, who had attacked the Communist doctrine of the dictatorship of the proletariat and subservience to Moscow at Tours, returned from prison camp in the summer of 1945 to ensure that, for the same reasons, the two groups stayed separate. The veteran Radical Édouard Herriot, who also returned from deportation to Germany, remembered the Popular Front rather than 1920, and made common cause with the Communists as head of the Unified Movement of the French Resistance. But he was outgunned by other party leaders who wanted to rebuild the Radical party as a bulwark against Communism. This was also the agenda of the small Union Démocratique et Socialiste de la Résistance (Democratic and Socialist Union of the Resistance, or UDSR), close to the Radicals, of which François Mitterrand was a founder member. The biggest new party of the Liberation, however, was the Mouvement Républicain Populaire (MRP), founded in November 1944. It was a Christian Democratic party which had its roots and values in the Resistance and which purged the incubus of the traditional association of Catholicism with the Right. It embraced the Republic, opposed large-scale capitalism, and believed in orderly and legal revolution. On the other hand, it was also opposed to collectivism—any attack on the principle of private property—and to Communism, and was supported by many on the Right who had no traditional party of the Right to vote for at the Liberation. De Gaulle, who was himself represented by no party, also looked to the MRP as the party that could best translate his views.

Battle was joined between de Gaulle and the political parties on the drafting and content of the new constitution. A referendum of 21 October 1945 voted by 96 per cent to reject the plan of the Radical party to restore the Chamber of Deputies and Senate of the Third Republic. It also accepted, by a smaller majority of 66 per cent, de

Gaulle's insistence that the Constituent Assembly should not be sovereign but should be limited in duration and powers and have its
constitution submitted to a referendum. However, elections to the
Constituent Assembly, held on the same day, returned the Communists
as the largest party. De Gaulle was elected head of the provisional
government unanimously by the Assembly on 13 November, but he was
forced to recognize the weight of the Communists by taking five of
them, including Thorez, as ministers when he formed his new government. As he had feared, the General found himself a prisoner of the
parties, especially the Communists and Socialists, and resigned on 20
January 1946.

De Gaulle fully expected that he would be recalled to office within a
week, on a tide of popular acclaim, and be able to dictate a constitution. Neither turned out to be the case. The Communists secured the
support of the SFIO for a constitution designed to reflect the sovereignty of the people in all its force and reminiscent of the Convention
of 1792: no upper house, and a weak executive. Unfortunately for
them, this was rejected by 10.5 million votes to 9.5 million in the

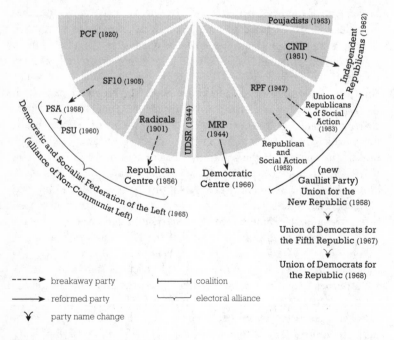

FIG. 2. Political Parties, 1944–1968.

referendum of 5 May 1946, and a second Constituent Assembly had to be elected. In this Assembly the MRP gained ground and the Radicals, now controlled by their right wing, returned from the dead. They were able to draft a constitution that included a second, indirectly elected chamber, the Council of the Republic, and a stronger executive. Despite ponderous speeches by de Gaulle warning against a parliamentary constitution, this was endorsed in a referendum of October 1946 by 9 million to 8 million, with 8 million abstentions. The socialist Vincent Auriol, who had been Blum's finance minister in the Popular Front government of 1936, was elected president of the Republic by the body of parliamentarians on 16 January 1947, while de Gaulle continued his twelve years of internal exile.

The parliamentary Republic

The Fourth Republic has never had a very good press. It presided over the decline of the French Empire, while its record of political crisis and ministerial instability—twenty-six governments between 1944 and 1958—was worse than that of the Third Republic. It was, however, a parliamentary regime constructed deliberately to ward off the twin evils of revolution and dictatorship, frequent guests at the feast of French history, and operated a system of coalition government which was not without its own rules and conventions.

A fundamental convention of the Fourth Republic was that only those who had gained some honour in the Resistance could legitimately hold office. This did not mean that ministers were plucked unwashed from the guerilla bands of mountain and forest, the *maquis*. Governmental competence was as important a qualification as political correctness. Thus the socialist Paul Ramadier, who became prime minister in January 1947, had not been in the underground Resistance but had voted against *pleins pouvoirs* for Pétain in 1940 and served on the liberation committee of Rodez in 1944. He had also been mayor of the mining town of Décazeville between 1919 and 1941 and after 1945, was under-secretary of state for mines between 1936 and 1938, and saw the bill nationalizing the gas and electricity industries through parliament in 1945. Neither did it mean that those who had been involved in the Vichy regime could not have their past rewritten by virtue of some 'deeds of Resistance'. Antoine Pinay, who became prime minister in 1952, had voted in favour of Pétain in 1940 and served on Vichy's national council. He had remained mayor of Saint-Chamond (Loire), where he had a leather business, until 1943, and thereafter been

president of Vichy's organization committee of the leather industry. He
was declared ineligible for elective office in 1945, but promptly managed
to have the ineligibility revoked on the grounds that he had worked for
the Resistance having false papers manufactured in his town hall. On
the basis of these 'deeds of Resistance' he was able to re-enter political
life and become premier.

The rise and fall of ministries, though by appearance anarchic, was
always dictated by certain rules. The prime minister was appointed by
the president of the Republic, not arbitrarily, but after extensive
discussions with the speakers of the National Assembly (Édouard
Herriot till 1953) and the Council of the Republic (Gaston Monnerville)
and party bosses. The prime minister then had to be invested by an
absolute majority of the National Assembly, according to the principle
of the sovereignty of the people, vested in its representatives. This was
the hurdle that provoked most of the crises of the Republic: 13 days in
August–September 1948, 35 days in May–June 1953, when *four* appointed prime ministers failed the investiture test, 36 days in September–November 1957. Once invested, the government depended for its
survival on its majority holding together; if one of parties forming the
majority deserted, the government would be forced to resign. The
power of party bosses, who were often not the same as ministers of that
party, was eloquent in this respect. Guy Mollet, for example, who kept
a bust of Robespierre on the desk of his town hall in Arras and was
elected secretary-general of the SFIO on a Marxist ticket in 1946,
forced Paul Ramadier to surrender office in November 1947. In this
respect, the Fourth Republic was the creature of party machines rather
than that of parliament itself.

For governments to stay in power, it was imperative to found them
on stable coalitions. Under the Fourth Republic, four durable coalitions were put together which ensured that, despite the succession of
individual ministries, the same group of ministers effectively held power
for long periods of time and the prime-ministerial office simply alter-
nated between the different party leaders. The main coalitions were
first, down to 1947, the Tripartism of PCF, SFIO, and MRP; second,
between 1947 and 1952 the Third Force of SFIO, MRP, and Radicals;
third, between 1952 and 1954, an alliance combining MRP, Radicals,
some Gaullists, and Independents (the Right); fourth, in 1956 and 1957,
a Republican Front dominated by Socialists and left-wing Radicals. In
1954–5 there was a curious interlude when Pierre Mendès France
attempted to break the tyranny of parties and party coalitions, while
after 1957 coalition government based on right-wing Radicals found it

increasingly difficult to sustain a majority between the Communists and the Right.

Tripartism was the rule from the first Constituent Assembly until the departure of Communist ministers from the government in May 1947. It bound together the three giant parties of the two Constituent Assemblies and the first legislature, elected in November 1946. But the presence of the Communists made coexistence difficult, especially when the Cold War began to bite. Guy Mollet believed that the only way to check a loss of voters to the PCF was for the Socialists to match the Marxism of the Communists, but Paul Ramadier looked to draw ever closer to the MRP. As shortages drove up prices and wages chased prices upwards, Ramadier imposed a policy of deflation which for a long time the Communist ministers accepted. But in April 1947 a strike broke out at the nationalized Renault car factory, citadel of the Communist-affiliated CGT. The Communist party put pressure on its ministers to agree wage rises; the ministers voted against Ramadier in a vote of confidence on 4 May but refused to resign. This was compounded by the snapping of the diplomatic link with the Soviet Union at the Moscow conference on 24 April and by France's urgent need for Marshall Aid. Ramadier duly dismissed the Communist ministers on 5 May.

In opposition, the Communists reverted to fomenting strikes and opposing colonial war. Their intransigence played into the hands of the Rassemblement du Peuple Français (RPF), launched by de Gaulle at Strasburg in April 1947. Given de Gaulle's opinions on parties, this was presented as the party that was not a party. It was supposed to be an extension of the wartime France Combattante, offering salvation from the 'degradation' of the country and its Empire where France Combattante had once ensured liberation. It had no programme but only certain themes: to reform French institutions, strengthen the Union, and restore French grandeur. It was a Rassemblement that invited membership from all parties or none, and it was possible to remain a member of another party while joining the movement. Thus Jacques Chaban-Delmas, who as 'national military delegate' had tried to bring COMAC into line with the Free French command in 1944, conquered the town hall of Bordeaux in October 1947 as a Radical, while also a member of the Rassemblement. Those municipal elections broke the mould of politics: the RPF won 38 per cent of the vote and took (as well as Bordeaux) Paris, Rennes, Lille, Nancy, Strasburg, and Marseille. The dominant parties of the regime felt threatened, banned double membership of their own party with the Rassemblement, and

formed a defensive coalition. Known as the Third Force, it linked the
SFIO, UDSR, MRP, and Radicals in the centre against RPF and the
Communists on the extremes, both seen in different ways as threats to
the constitution.

If any one prime minister typified the Third Force, it was Henri
Queuille. A country doctor, he was a Radical with his fief in the
Corrèze, where he was mayor of Neuvic d'Ussel between 1912 and
1965, and member of the departmental *conseil général* between 1913
and 1961. An expert on peasant questions, he was minister of agriculture in eleven governments between 1924 and 1940. He abstained in the
vote giving full powers to Pétain and joined the French Committee of
National Liberation in London in 1943. He rebuilt the Radical party
after its electoral disaster of 1945 and was continuously in office
between 1948 and 1954. He was four times prime minister in that time,
the longest, between September 1948 and October 1949, something
of a record. His skill was to bind together a coalition which stretched
from the SFIO to the Independents, and which previously had been
torn apart by a continuation of pre-war conflicts between the
socialist Léon Blum and the right-winger Paul Reynaud. During his last
premiership, in May 1951, Henri Queuille sponsored an electoral
law—the so-called *loi des apparentements*—specifically designed to
ensure the perpetuation of the Third Force against Gaullist and
Communist opposition. This ruled that, under the system of departmental proportional representation that obtained, joint lists of parties
would take all the seats in a department if they secured an absolute
majority. The Third Force parties played the game successfully, and
took 283 seats. De Gaulle, on the other hand, who wanted the RPF
to emerge as a majority party with its hands united, refused to allow
such pacts, so that the RPF secured only 120 seats instead of an
anticipated 200.

The elections of June 1951 nevertheless transformed the pattern of
politics in France. It saw a return of the parties of the Right, regrouped
as the Conseil National des Indépendants et Paysans (National Council
of Independents and Peasants, or CNIP). The RPF was now the largest
party, and de Gaulle demanded the right to form a government,
insisting at the same time that he would change the constitution.
President Auriol refused, saying that he had no wish to play a French
Hindenburg. But the situation on the Left was no more inspiring. There
were still over 100 Communists, who were not permitted to form an
element of any majority. The SFIO, meanwhile, refused to back any
governments that were based on any right-wing support, and for the

first time since 1945 left the ruling majority. Auriol's room for manœuvre was thus extremely tight. 'The only possibility at the moment', he said in January 1952, 'is a centrist majority without the socialists and with the RPF, or with the socialists and without the RPF, but 40 Independents would still be required. This Assembly is impossible! It has no civic sense!'[4]

In the end Auriol opted for a system that left the Socialists on one side, used the right-wing Radicals and MRP as the base, and split the RPF by inducing some of them, to the fury of de Gaulle, to support the government majority. Antonie Pinay was invested on 6 March 1952 thanks to the support of 27 of them, who now called themselves the Action Républicaine et Sociale (Republican and Social Action) group. René Mayer, a right-wing Radical, secured the support of all the RPF deputies for his ministry in January 1953 in return for a commitment to constitutional reform. This treachery, combined with the poor performance of the RPF in the municipal elections of May 1953, provoked de Gaulle to withdraw the whip from the RPF on the grounds that by accepting the 'games, poisons and delights of the system' they had legitimated them. Henceforth he looked forward to a 'great upheaval' that would destroy the system itself.[5] For their part the former RPF deputies, now rebaptized the Union des Républicains d'Action Sociale (Union of Republicans of Social Action) were happy to be part of the system. In June 1953 they supported the investiture of the Independent Joseph Laniel, a Normandy textile industrialist who had served in the Paul Reynaud government of March 1940 and had voted full powers to Pétain in July 1940, but had been invited onto the National Resistance Council in 1943 in order to counterbalance the Communists.

The 'system' under fire

The system reached a low point with the election of the new president of the Republic in December 1953, as Vincent Auriol came to the end of his seven-year term. The main problem was that there were no inspiring candidates in contention. Such was the bankruptcy that many wished the 81-year-old Édouard Herriot, recently acclaimed as the 'Republic in person', to stand, but he declined on the grounds of ill health. The parliamentarians held thirteen ballots over seven days, and eventually settled on a compromise candidate, the Independent René Coty, advocate of the business interests of Le Havre, vice-president of the Conseil de la République, with his ruddy complexion and pin-striped suit. Auriol, receiving his successor on Boxing Day 1953, was manifestly not impressed.

In 1954–5 there was a brief attempt to reform the system by the brilliant but maverick politician Pierre Mendès France. A Radical, he was not part of the Radical establishment but a Young Turk of the 1930s who was now supported by the dynamic Jacobin Club, formed by angry young Radicals in 1951, and by the political weekly of Jean-Jacques Servan-Schreiber, *L'Express*. 'Listen to the murmurs of discontent rising,' he told the Radical party conference in November 1953. 'We are in 1788!'[6] Mendès France was against the coalitions of parties, interest groups, and class politics that were dividing the nation and tearing apart the state. Inspired by Gambetta and Clemenceau, he wanted to strengthen the authority of the republican state, which would articulate the general will of the sovereign people and unflinchingly pursue the common good. He spoke directly to the people through the Saturday evening radio broadcasts he instituted, and believed himself to be in touch with them through the mail he received. When he formed his government he split parties down the middle, and was supported by half of the Radicals, SFIO, UDSR (Mitterrand was his minister of the interior), and former RPF. On the other hand, he was wholly opposed by the Communists, MRP, and the Independents. His programme was innovative: to get rid of the burdens of empire and to invest in economic modernization, increase the standard of living, and attack social injustice.

Mendès France took on the party system, beginning with his own Radical party. He was opposed by the Radical bosses such as René Mayer and Henri Queuille, who helped to topple him from power in February 1955. He was replaced by one of the Radical party bosses, Edgar Faure, a barrister with his fief in the Jura, who formed a government with the support of the MRP and the Independents. Mendès France had his revenge at a specially convened Radical congress in May 1955. While Faure argued that the party should stay in power and form coalitions with parties to the Right, Mendès France argued that the party should have a clear programme like the New Deal and be 'a party of renewal and opposition'. He won the debate and conquered the leadership of the Radical party. He also took his revenge on Faure after the latter dissolved the National Assembly ahead of the expiry of its mandate, in the hope of winning a snap election. Although this was within the letter of the constitution, it was against the spirit, certainly for the Radical party, which saw itself as the heir to Gambetta and his stand against the authoritarian dissolution of the Chambre des Députés in 1877. Mendès France managed to get Faure expelled from the party. The bosses made a final attempt to unseat Mendès France at the radical conference of October 1956; failing, they went off to form a separate party, the Centre Républicain. Mendès France had had his way, but he effectively destroyed the Radical party in the process.

Mendès France was an enemy of the party system, but believed in articulating the general will of the sovereign people. In this respect, though he opposed de Gaulle, the resemblance was intriguing. Unfortunately, the will of the people to which he paid homage was not always as general and enlightened as he would have wished. Neither was it a Rousseauist abstraction, but a body of angry *sans-culottes* protesting against the recession and tax increases. And far from idealizing the republican state they had a peculiar dislike of moralizing, left-wing, Jewish politicians like him.

The period of Mendès France's premiership saw the rise of the Poujadist movement. Known officially as the Union for the Defence of Shopkeepers and Artisans, it was launched in July 1953 by Pierre Poujade, a bookseller and newsagent of the small town of Saint-Céré in the Lot. It protested against tax increases and tax inspections by the Revenue at a time of economic downturn and long-term rural depopulation that threatened the businesses of those who served the farming community. Poujade was a red-blooded local demagogue, one of seven children and himself married to a *pied-noir*, the head of a large family. The death of his father in the First World War had forced him to leave school early to earn a living, and he had served in the air force during the Second. He was not impressed by local politicians such as Gaston Monnerville, mayor of Saint-Céré, president of the *conseil général* of the Lot, speaker of the Council of the Republic—and a black from French Guiana. Neither was he impressed by Mendès France, whom he accused of having not 'a drop of Gallic blood' in his veins, of planning to cover France with supermarkets, and of insulting French wine-growers and café-owners by launching a campaign against alcoholism and drinking glasses of milk at international conferences. Indeed, Pierre Poujade was not impressed by anyone—politicians, intellectuals, bureaucrats, Eurocrats, plutocrats, technocrats—and believed in taking matters into his own hands.

By the end of 1953 his movement had taken over the Lot chamber of commerce and held its first departmental congress in Cahors. In November 1954 it held its first national congress in Algiers, and in January 1955 it organized a rally in Paris, attended by 150,000.

In March 1955 it invaded the public galleries of the National Assembly in an attempt to stop tax increases. The roller-coaster was in fact full of contradictions. It was for some time supported by the Communist party and used the rhetoric of the Left, Poujade (loosely) citing the Declaration of the Rights of Man of 1793 that 'When the government violates rights guaranteed by the constitution, Resistance in every form is the most sacred of rights and most imperious of duties'.[7] On the other hand the movement was fiercely nationalist, using a Gallic cock as its symbol, joined by former paratroopers of the Indo-Chinese war and militants of the extreme Right like Jean-Marie Le Pen and campaigning for the defence of French Algeria; Poujade himself was nicknamed 'Poujadolf'. It was a political movement that refused to become a party, lest it become part of the 'system' it despised, and found a clever formula in the summer of 1955 by campaigning for an Estates General to consider *cahiers de doléances*, or registers of grievances. But thrown a challenge by Faure's dissolution of the National Assembly in November 1955, it formed a party and won 12 per cent of the vote and 52 seats.

The Poujadist success in the election of 2 January 1956 split the Right and deprived it of the advantage it had held since 1951. It was a victory for the Republican Front of Mendès France Radicals and the SFIO under Guy Mollet, the title of which, and the logo—a Phrygian bonnet—had been thought up by *L'Express*. The victory, however, was limited: they had only 28 per cent of the vote and 170 seats instead of the 300 they had hoped for. To the Left were the Communists with 146 seats, eager to revive the Popular Front of 1936 and become part of the governing coalition for the first time since 1947. Guy Mollet, who was asked to form a government by René Coty in preference to Mendès France, feared that the PCF wanted to seduce voters away from the SFIO and firmly rejected the idea of a Popular Front. Besides, in spite of Khrushchev's 'secret speech' to the 20th congress of the Soviet Communist party, Maurice Thorez kept the PCF committed to the traditional doctrines of violent revolution and dictatorship of the proletariat, and endorsed the Soviet invasion of Hungary in November 1956. Opposed by the Right, Mollet was obliged to turn to the Radicals around Edgar Faure and to the MRP in order to secure investment on 5 February. At once, the Mollet government found itself embroiled in the Algerian war and following a strategy of colonialist repression. This caused Mendès France to resign in May 1956 and alienated a whole section of the SFIO who argued that socialist principles were being betrayed; in 1958 they broke away to form the Parti Socialiste Autonome.

In May 1957 the Mollet government was overthrown by an unholy alliance of the PCF and the Right. The Republic was now almost ungovernable. This was underlined by the ministerial crisis from 30 October to 5 November 1957, when five successive premiers were designated, two of whom (Mollet and Pinay) got as far as to request investiture attempts by the Assembly, only to be rejected. Ironically, Félix Gaillard, formerly the head of Jean Monnet's private office and rising star of the anti-Mendès France Radicals, invested on 6 November 1957 at the age of 38, the youngest French head of government since Napoleon Bonaparte, lasted only six months in power. Decisions about the future of the Republic were then being made not in Paris but in Algiers.

A *coup d'état* or not a *coup d'état?*

When the military and right-wing extremists seized power in Algiers on 13 May 1958 and called upon General de Gaulle to form a government of public safety it was not, paradoxically, a moment of ministerial crisis. It was, rather, the day of the investiture of Pierre Pflimlin, an Alsatian MRP politician who had served both Vichy and the Fourth Republic as an examining magistrate. He asked the Assembly to give him three months to revise the constitution, but he was also seen to favour negotiation with the Algerian nationalists, and it was to prevent his investiture that, the coup was launched.

In the short run the coup had the effect of strengthening his hand. The SFIO rallied to him in the investiture vote and Guy Mollet was taken on as vice-president of the council. General Salan, the commander-in-chief in Algiers, was telephoned and given civilian powers as governor to restore order. Unfortunately, on 15 May Salan bowed to the coup leaders and shouted 'Vive de Gaulle!' from the balcony of the governor's building in the forum of Algiers. De Gaulle replied by releasing a communiqué to say that was 'ready to assume the powers of the Republic'. This was the first move in a complicated and clever double game to return to power. He used the threat of a military coup while affecting to have nothing to do with it, and insisted that if he returned to power it would be by legal means, through a delegation of exceptional powers by parliament. 'Do people believe that at the age of 67 I am going to begin a career as a dictator?' he asked a crowded press conference on 19 May, blatantly overlooking the fact that Marshal Pétain had begun his career as a dictator at 84.[8] De Gaulle was adamant that there should be no repeat of 1946 or 1951: if he were to return to power, it must be to change the constitution, not to find himself part of the system. So he let the crisis spin out in order to present himself, as in 1940, the saviour of the nation. To the military and extremists in Algiers he pretended that he was their man, ready to set up a strong government to save French Algeria, while to the political class in Paris he pretended that he was the sole guarantor of the liberties of the Republic against military dictatorship and fascism.

During the night of 26–7 May de Gaulle visited Pflimlin and tried to bully him into resigning. Pflimlin held firm, and the next day his plans for constitutional revision were approved by a vote of confidence in the National Assembly. In Algiers General Massu threatened to put into action Operation Resurrection: a seizure of government buildings in Paris by paratroopers. A demonstration of 200,000 people took place in Paris on 28 May, orchestrated by the Communist party, which argued that, far from being a barrier to 'military and fascist dictatorship', de Gaulle was emerging as a new Pétain or Louis-Napoleon Bonaparte.[9] At this moment leading politicians such as Mollet, Auriol, and Pinay deserted Pflimlin and rallied to de Gaulle. Between the peril of military dictatorship and the peril of Communist insurrection, the General seemed the only escape. Pflimlin resigned, and de Gaulle was invited by President Coty on 29 May to form a new government. On 1 June de Gaulle asked the Assembly for full powers to restore order and unity, draft a new constitution, and submit it to the people for ratification. The parties, in a state of panic, split down the middle and invested him by 329 votes to 224.

Having learned the lessons of 1946 and 1951, de Gaulle adopted a measured approach, in order to belie accusations, being made by the Communists, some socialists, and Mendès France, that he was himself embarking on dictatorship. For the time being he was the last prime minister of the Fourth Republic, and conducted business at Matignon, while Coty remained at the Élysée. Though he had repeatedly criticized the 'exclusive regime of parties', he offered posts in his government to leaders of all the leading tendencies: Mollet for the SFIO, Pinay for the Independents, even poor Pflimlin for the MRP, together with Houphouët-Boigny of the African Democratic Rally to oversee the transition of the French Union into the Community. The maverick Jacques Soustelle, ally from the days of the Resistance and the RPF, was not brought in as minister of information until July. De Gaulle was clear that there would be no Constituent Assembly. The constitution was drafted by a committee of ministers under his chairmanship and a committee of jurists, including members of the Conseil d'État, which offered governments advice on legislation, under Michel Debré, now minister of justice. It proposed a significant shift of power from the parliament to the president. The president could appoint the prime minister, dissolve the National Assembly once a year, refer legislation to a Constitutional Council if the Assembly was thought to have exceeded its powers, hold referenda on important constitutional matters, and take emergency powers in a state of national crisis. He was elected indirectly by an electoral college of 80,000, composed of elected representatives of all sorts. The parliament's powers to interpellate the government, pass a vote of censure against a prime minister, and amend legislation, were reduced, and ministers had to resign their parliamentary seats.

De Gaulle underlined his republican credentials by revealing the constitution on the place de la République on 4 September 1958, anniversary of the declaration of the Third Republic in 1870. There was then a change of gear. All the tricks of the media were used by Soustelle in the run-up to the referendum on 28 September, when the constitution was endorsed by 79 per cent of voters. A new official party, the Union for the New Republic (UNR), was founded by Soustelle for the parliamentary elections of 23/30 November 1958. It emerged as the largest party, with 20 per cent of the vote and 198 seats. The Independents did well and the MRP held up but the Communists, SFIO, Radicals, and Poujadists all collapsed. There was the biggest turnover of deputies since 1919, with only 131 of the 537 sitting deputies returned. De Gaulle was elected president of the Fifth Republic on 21 December, against unfavoured Communist and Socialist rivals, with 78 per cent of the college vote. On the day of his inauguration, 8 January 1959, he laid a wreath at the Tomb of the Unknown Warrior at the Étoile with outgoing president René Coty, and then drove back down the Champs-Élysées, leaving Coty, totally bewildered, on the pavement. Thus the Fourth Republic faded into history.

Towards a republican dictatorship

Once de Gaulle was safely established in power, the façade of caution soon fell away to reveal his real intention to establish a highly personal and presidential regime. This was betrayed not only in the actual workings of power but in the General's own utterances. On 4 September 1958 he had called himself a 'national arbiter' to see fair play between the different branches of the constitution, even though he was himself the leading player. In December 1958 he proclaimed that as 'Guide of France and head of the republican state, I will exercise supreme power over the whole range that it now encompasses and according to the new spirit that entrusted it to me'. The increasing mysticism of his language was revealed when he addressed the nation during the 'week of the barricades' in Algiers in January 1960 and spoke of 'the national legitimacy that I have embodied for twenty years', as if the authority to speak for France he had asserted on 18 June 1940 had never been interrupted, either by the delegation of full powers to Pétain or by the constitution of the Fourth Republic. The hubris of his declarations reached their apogee at a press conference in January 1964 when, finally abandoning the fiction of the separation of powers, de Gaulle announced that 'the indivisible authority of the state is entrusted wholly to the president by the people who have elected him.'[10]

The development of de Gaulle's powers took place both *vis-à-vis* parliament and *vis-à-vis* the government. In January 1959 he appointed Michel Debré prime minister. Debré served with devotion and loyalty. He shared the General's view of the unity of the executive, that no daylight should be seen between the president and the government, and accepted the emerging doctrine of the 'reserved powers' of the president in matters of defence, foreign policy, and the Community, because like de Gaulle he believed in the independence and greatness of France. Even so, de Gaulle did not make things easy for Debré. His private office, ostensibly to deal with the 'reserved powers', soon trespassed on the domains of other ministries. Moreover, there was very little idea of collective ministerial responsibility: ministers were hired and fired by de Gaulle, were consulted in private behind Debré's back, and were not allowed, in rare councils of ministers, to express opinions on matters outside the brief of their departments.

De Gaulle cut ministers off from their parliamentary power base, and freed his hands from having to construct ministries reflecting the balance of power in the Assembly, by forcing them to give up their parliamentary seats. Guy Mollet and the socialists refused to serve in Debré's cabinet and Antoine Pinay was sacked for speaking out of turn. The General also appointed to the government a significant number of technocrats, like Wilfrid Baumgartner, governor of the Banque de France, who replaced Pinay at the finance ministry, and Maurice Couve de Murville, ambassador in Bonn, who became foreign minister. Parliament itself was supposed to cause no trouble. Debré told it in January 1959 that 'the depoliticization of matters of national importance is a major imperative'.[11] His government was duly invested by 459 votes to 56. Jacques Chaban-Delmas, a collaborator of de Gaulle from the days of the Resistance and the RPF, became speaker of the National Assembly. The Union pour la Nouvelle République, which threatened under Jacques Soustelle to become a vehicle of partisans of French Algeria, was brought to heel as a party of government after Soustelle was expelled from both government and party in 1960, for defending the insurgents of the 'week of the barricades'. The government exploited this crisis to obtain full powers from parliament to legislate by ordinance for a year after February 1960. But when, the following month, a majority of deputies signed a petition requesting the emergency recall of parliament to discuss an agricultural crisis, de Gaulle blatantly flouted the constitution by turning them down. On the other hand, he was only too keen to appeal over the head of parliament to consult the people by referendum on the fate of Algeria.

While the Algerian war lasted, de Gaulle benefited from what amounted to a state of emergency. 'As soon as the peace is concluded', he concluded '[the parties] will try to get rid of me. At that moment I shall attack.' His method of attack was to restore the direct election of the presidency of the Republic. Direct elections had been eschewed by the Third, Fourth, and (until now) the Fifth Republics because the last time they had been tried, in 1848, a landslide had been won by Louis-Napoleon Bonaparte, who went on to destroy the Second Republic and make himself the Emperor Napoleon III. Michel Debré, though he had argued in 1945 in favour of a 'republican monarch', opposed direct elections on the grounds that they would strengthen the president even more and erode what little autonomy the government retained. He was removed from the premiership and replaced by Georges Pompidou, who had been deputy director of de Gaulle's private office in 1944–6 and its head in 1948–53 and 1958–9. Whereas Debré had been a member of the Council of the Republic throughout the Fourth Republic, Pompidou had never held elective office, having been a schoolmaster until the Liberation, then a director of the Rothschild bank. As such, he could be expected to be de Gaulle's poodle.

The National Assembly dealt appropriately with such a snub. Radicals and socialists refused to serve in the government, which was invested on 26 April 1962 by an unimpressive majority. Then, on 15 May, the MRP ministers left the government in protest at an anti-European press conference held by de Gaulle. This in no sense deflected the General from his purpose. Exploiting the sympathy gained by surviving an OAS assassination attempt, he announced on 20 September that there would be a referendum on the direct election of the president, arguing that this would give the president the greater powers required for the strength and continuity of the Republic. The outburst from the political class was immediate and vociferous. Communists and Socialists brandished the spectre of de Gaulle pulling on the boots of Napoleon III or restoring absolute monarchy. Guardians of the constitution pointed out that the president could not make this change without the consent of both houses of parliament. Gaston Monnerville, speaker of the Senate as he had been of the Council of the Republic, told the Radical congress in September that 'in my opinion a motion of censure is the direct, legal and constitutional reply to what I call an abuse of authority.'[12] In this opinion he was supported by the jurists of the Conseil d'État and the retired statesmen of the Conseil Constitutionnel, although the latter in fact had no jurisdiction over the actions of the president. De Gaulle, for his part, was ready to take on the parties and the politicians and battle to the death.

A motion of censure against Pompidou was voted on 5 October 1962 by 280 votes out of 480, Independents and MRP joining SFIO and Communists. De Gaulle decided to keep on his prime minister and dissolved the Assembly instead. A 'cartel des non' was formed by the politicians to defeat de Gaulle in the referendum on 28 October. 'From the socialists to the Independents on the Right,' said Michel Debré, 'the former tenors of the Fourth Republic rivalled the Communists in the violence of their attacks.'[13] It was a close-run thing: both Mollet and Monnerville fully expected to be forming an interim government. In the event, de Gaulle managed to secure 62 per cent of the vote for his reform. In the second round, the parliamentary elections of 18/25 November, de Gaulle took on the parties themselves. The result was a triumph for the UNR, which secured 35 per cent of the vote and 233 seats, nine short of an absolute majority. The SFIO and the Communists held up but the MRP lost ground and the Radicals and Independents were shattered, the latter dropping from 103 seats to 29, and neither large enough to form a group in the Assembly. Out of the rubble of the Independents Valéry Giscard d'Estaing—who had passed out second from Polytechnique and third from the École Nationale d'Administration, the two grandest of the *grandes écoles*, married into the Schneider steel fortune, and inherited the Auvergne seat of his Pétainist grandfather in 1956—forged a group of Independent Republicans which completed the government majority and was taken on, at the age of 35, as Pompidou's finance minister.

A new public sphere

The triumph of the Gaullist regime was the failure of the political parties. In spite of their principles they had rushed to de Gaulle as a saviour, only to find themselves reduced to impotence by him. Oblivious, the party bosses clung on, negotiating pacts and deals, blocking any new ideas or initiatives. Some broke away from old parties to found new ones: such was the Parti Socialiste Autonome (Autonomous Socialist Party), which separated from the SFIO in September 1958 and was renamed the Parti Socialiste Unifié (Unified Socialist Party, or PSU) in 1960. This argued that the compromising of socialist principles begun by Mollet over the Algerian war had been completed by his endorsement of de Gaulle's coup. It claimed the humanist inheritance of Blum and Jaurès and won over Pierre Mendès France as a member.

Beyond the realm of the political parties, old or new, however, there was opening up a new political sphere. It was defined by individuals who had never belonged to political parties, or who had left in disgust at the combination of doctrinaire ideologies and sordid machine politics, or remained within political parties but sought an alternative power-base to the party machines. The rank and file were made up of the new middle classes—executives and technocrats, teachers and lecturers, civil servants, and trade union officials—whose weight in society had grown as a result of economic expansion and modernization since 1945. They regarded the old parties with scorn, and were in search of a new politics to deal with the problems and challenges of rapid social change. They espoused managerial values of efficiency and modernization, but were not happy with the depoliticization imposed by the Gaullist regime, and sought to define a new public sphere for civic responsibility and political participation. Their viewpoints were articulated by political weeklies such as *L'Express* or *Le Nouvel Observateur*. The main instrument of their participation, however, was the political clubs which sprang up in the 1960s as *sociétés de pensée* had during the Enlightenment two centuries before. They included the Club Jean Moulin, founded by Daniel Cordier, the secretary of the Resistance leader Jean Moulin when he was prefect of Chartres in 1940, the Cercle Tocqueville, Citoyens 60, the Club des Jacobins, founded as early as 1951 by the Radical Charles Hernu, and the Ligue pour le Combat Républicain, founded in 1959 by François Mitterrand, for whom the coup of May 1958 was an attack on republican legality and who from now on redefined himself increasingly as a Socialist. These last two clubs formed the nucleus in June 1964 of the Convention des

Institutions Républicaines (Convention of Republican Institutions, or CIR).

Though they tried to remain independent of party politics, the political clubs were nothing if not political, and the election of a president by direct universal suffrage, due in December 1965, was a first challenge. Influenced by the 'making' of John F. Kennedy in the United States, *L'Express* projected an identikit portrait of an ideal president in September 1963, 'Monsieur X'. He would be a manager who could rise above sects and even ideology, at ease with the new scientific and technological discourse, able to relate to emerging interest groups in society, such as trade unions and students. The preferred candidate of *L'Express* and clubs such as the Club Jean Moulin was Gaston Defferre, mayor of Marseille and architect of colonial reform. It soon became clear, however, that such a candidature could not do without the endorsement of political parties. A Socialist, Defferre was endorsed in December 1963 by the SFIO, but rejected by the PSU, MRP, and Radicals. Needing a broader base, he tried to forge a 'grand federation' in 1965, gathering support from the SFIO, MRP, and Radicals. The MRP leader, Jean Lecanuet, prohibited him from using the term 'socialist' in his manifesto, while the SFIO boss Guy Mollet condemned his shift to the Right. Between them they torpedoed his chances, and in June 1965 he withdrew.

François Mitterrand then entered the lists. His advantage over Defferre was, first, that he rejected centrist politics and embraced an alliance with the Communists to become the single candidate of the Left and, second, that he was able to impose himself on the Socialist and Communist party bosses by dint of having a power-base in the CIR. Now that the centre ground was vacated, Jean Lecanuet stood for election, supported by the MRP, Radicals, and Independents. De Gaulle, exploiting his presidential prestige, refused to announce his candidature until a month before the first ballot and then declined to campaign. Panic swept his camp as public-opinion polls showed his predicted score slipping from 66 per cent to 46 per cent, and in the event he secured a mere 43 per cent, against 32 per cent for Mitterrand and 16 per cent for Lecanuet. This meant the humiliation of being forced into a run-off against Mitterrand. At this stage he did consent to appear on television, interviewed by the complaisant journalist Michel Droit, dismissing Mitterrand as the candidate of the political parties while he was the candidate of History. In the second ballot, on 19 December 1965, he secured 55 per cent of the vote against Mitterrand's 45 per cent.

De Gaulle's re-election, which should have consolidated his power, in fact revealed increasingly worrying cracks in the edifice. He had been forced into a political dog-fight and to reveal himself as partisan, no longer the president by universal acclaim. For the parliamentary elections of March 1967, Pompidou tried to forge a new party of the majority that would run only one candidate in each constituency, the Union des Démocrates pour la Vᵉ (Union of Democrats for the Fifth Republic, or UDVᵉ). But Giscard d'Estaing, who had been dismissed as finance minister in 1966 for his unpopular deflation policy and was already thinking about politics after de Gaulle, presented his Independent Republicans as liberal, European, and centrist, and when asked whether he supported de Gaulle conceded 'oui, mais'. Meanwhile Mitterrand had transformed his CIR into a Fédération de la Gauche Démocratique et Socialiste (Democratic and Socialist Federation of the Left, or FGDS), incorporating the non-Communist Left including the SFIO, and made an electoral pact with the Communists. Accordingly, the Left recovered well in the elections, the Centre Démocrate (Democratic Centre), formed by Lecanuet from the MRP, Independents, and some Radicals, was squeezed, while the UDVᵉ, winning 37 per cent of the popular vote, took only 200 seats and was in the invidious position of having to rely on the 45 Independent Republicans for a majority. De Gaulle would have replaced Pompidou as prime minister had Pompidou's chosen successor, the stony-faced foreign minister, Couve de Murville, not lost his seat. Pompidou struggled on, harried by Mitterrand on one side and Giscard on the other, obliged to resort to ruling without parliament and legislating by ordinance.

The revolution of May 1968

From one perspective, it seemed that the opposition was doing well. But from the point of view of discontented forces in French society, the political parties had recovered some initiative while still being unable to find a way through. The FGDS had reverted to the practice of electoral pacts, but had still only won 121 seats. The Communists had recovered well, but still maintained a Stalinist inflexibility, and in 1965 had expelled Trotskyists led by Alain Krivine from the Union of French Communist Students. The Unified Socialist Party (PSU), which had broken from the SFIO over the Algerian war, was open to new ideas, and was led from 1967 by Michel Rocard, remained no more than a sect. The National Union of French Students (UNEF), which had boasted 100,000 members at the time of the Algerian war, now had no more than 50,000 and confined itself to corporate problems of students.

The discontent that was fermenting in the student body thus has no effective outlet in the existing parties or unions. Universities had been expanding in recent years. The arts and social science faculties of Paris university had been moved to an overspill site on the former shanty town of Nanterre in 1964, and had grown from 4,000 to 15,000 students by the autumn of 1967. The university world reflected in microcosm the authoritarianism, hierarchy, and bureaucracy of the Gaullist state: no representation of students, little dialogue between teachers and students, the dead hand of structuralism that allowed no place for individual creativity, and strict separation of the sexes in the accommodation blocks. Expanding education was geared to forced economic growth, new technology, higher productivity, and the consumer society, and students were required to adapt to this as white-collar workers in the public or private sector. The outside world, meanwhile, was exploding, from the war in Vietnam and the Cultural Revolution in China to the Latin American revolutions of Fidel Castro and Che Guevara, and from the black civil rights movement in the United States to the Prague Spring. This spilled over into the campuses of Europe, and an international student demonstration against the Vietnam war in Berlin disgorged activists such as the Trotskyist Alain Krivine and the anarchist Danny Cohn-Bendit, who carried the struggle back to Nanterre in January 1968.

The authorities reacted to the disturbances with a fatal combination of repression and weakness. Militants were arrested at Nanterre on 20–1 March 1968. This provoked the occupation of the administrative block at Nanterre by Cohn-Bendit and his comrades on 22 March. At this stage the student union under its president Jacques Sauvageot and the lecturers' union (SNE-Sup) under Alain Geismar began to mobilize. Teaching was suspended at Nanterre on 26 March and the whole campus closed on 2 May. The result was that agitation immediately switched to the Sorbonne. The decisive moment was the afternoon of 3 May 1968, when the police were ordered into the Sorbonne to arrest political and union leaders and the university was closed. The lecturers' union declared a strike in solidarity with the students. Demonstrations spread to the Latin quarter, and school students became involved. Repression by the police was stepped up, and during the night of Friday–Saturday 10–11 May barricades were thrown up in the streets of Paris.

The government vacillated. Pompidou was away visiting Iran and Afghanistan between 2 and 11 May. De Gaulle wanted to send in the army early on 11 May, but his army minister Pierre Messmer warned him that the conscripts might well fraternize with the students. Back in

Paris that evening, Pompidou decided that the only way to avoid student deaths was to reopen the Sorbonne. De Gaulle opposed him, the education minister resigned, but Pompidou had his way. The Sorbonne now became a student commune, occupied by the students who turned it into a libertarian utopia, holding endless meetings, covering the walls with graffiti, questioning anything and everything about society. The Communist CGT and non-Communist CFDT and Force Ouvrière called a one-day strike on Monday 13 May. In Paris 800,000 demonstrators marched behind banners declaring the solidarity of students, teachers, and workers, while slogans like 'Dix ans, ça suffit!' and 'De Gaulle, au musée!' were chanted.

De Gaulle seemed to get the message and on 14 May flew off on a state visit to Romania. Television viewers saw de Gaulle watching Romanian folk dancers while France hovered on the brink of chaos. For the general strike triggered a spontaneous strike wave among workers, who themselves were suffering from forced productivity increases according to the new management gospel of Taylorism and a total lack of consultation in the workplace. Action was taken on the shop floor, bypassing union bureaucracies. Though the Communist Georges Marchais called Cohn-Bendit a 'German anarchist' and the CGT leader Georges Séguy refused to shake his hand on 13 May, young workers especially felt solidarity with the students who were in the front line against baton charges and water cannon. The strike spread from the aircraft and automobile industries to the railways and metro, then to the electronic and engineering industries and to department stores and the civil service. Before long 10 million people were on strike. New demands were being made, not only for better wages and conditions, but for workers' control or *autogestion*.

De Gaulle returned late on 18 May, and on 24 May broadcast to the nation his intention of holding a referendum on 13 June on greater participation in universities and industry. This time his appeal to the people fell flat. Fortunately for him, the CGT and Communist party, far from seeking to whip up revolution, were trying to direct the spontaneous movement of strikes and demonstrations into conventional trade union and political channels. They wanted to regain control of the situation for the union and party, limit the workers' demands to pay and conditions, and force concessions on these from employers and the government. Pompidou saw the need to do business and, flanked by Édouard Balladur from his private office and the employment minister Jacques Chirac, he brought union and employer representatives together at the ministry of labour, rue de Grenelle, and negotiated the Grenelle agreements. Though these should have bought off the workers, they did not, and Georges Séguy was shouted down when he announced them to the Renault car workers at Billancourt.

At this point the state was in crisis. Though the CGT and PCF did not seem to be a problem, other unions and politicians were preparing to step into the breach if de Gaulle lost the referendum. The student union, the CFDT, which had taken up the idea of *autogestion*, and the PSU, radicalized by an influx of students, held a rally of 60,000 at the Charléty stadium on 27 May at which Mendès France was produced as a possible candidate to head an interim government. François Mitterrand held a press conference on 28 May to declare himself ready to head an interim government candidate and run for the presidency. Not to be outdone, the Communists now declared that they would hold a mass demonstration in Paris on 29 May in support of their demand for a 'popular government of democratic union', including themselves.

Pompidou ordered tanks to be brought up to the outskirts of Paris and warned de Gaulle that the Communists might attempt another Paris Commune. Then, on 29 May, de Gaulle boarded a helicopter and disappeared. It was thought that he might have gone to Colombey but he had not. That evening it emerged that he had been to Baden to meet the commander-in-chief of French forces in Germany, General Massu. His plans were unclear. One hypothesis was that he was suffering depression and about to resign, and that Massu persuaded him out of it. Another was that, as in May 1958, he was looking to Massu's forces to solve the crisis. A third was that, simply by disappearing, he hoped to bring the country to its senses. 'I want to plunge the French people, including the government, into doubt and anxiety,' he told his son-in-law, Alain de Boissieu, 'in order to regain control of the situation.'[14]

It seemed to work, at least in the short run. De Gaulle returned to Paris and was acclaimed by a demonstration of 500,000 loyalists on the Champs Élysées. He recovered his oratorical skills and appeared on television to warn the country of the threat of Communist dictatorship. Pompidou persuaded him to cancel the referendum and hold elections instead, scheduled for 23 and 30 June. In the first ballot the silent majority, who had gazed into the void, gave 46 per cent of the vote to the Gaullist party, refashioned for the occasion as the Union des Démocrates pour la République (Union of Democrats for the Republic, or UDR). In the second round the Gaullists consolidated their victory, securing an absolute majority. In the longer run, however, de Gaulle's days were numbered. To see him swept away by revolution was frightening, but once the Gaullists had secured a firm grip on the state, they could think seriously of a peaceful succession.

'General de Gaulle? he no longer exists,' reflected Pompidou on 11 May. De Gaulle for his part criticized the way Pompidou had handled the crisis, particularly the reopening of the Sorbonne and the Grenelle agreements, and this he confirmed by dismissing Pompidou in July and replacing him by Couve de Murville. Pompidou now became the darling of the conservative Gaullists, as the General pressed ahead with a far-reaching university reform. De Gaulle, ever one for dramatic gestures, resurrected his referendum and pinned it to a reform of regional government and the Senate, which had not been forgiven for its hostility in 1962. This was presented as a vote of confidence in de Gaulle, and his rivals and enemies now saw the chance to be rid of him safely. In the referendum campaign he was thus opposed not only by the Left, Lecanuet's Centre Démocrate, and the Radicals but also by Alain Poher, who had recently taken over as speaker of the Senate from Gaston Monnerville, and decisively by Giscard d'Estaing of the Independent Republicans, who late in the day urged a 'no' vote. In the referendum of 27 April 1969 de Gaulle was rejected by 53 per cent of the vote to 47 per cent. True to himself, he respected the verdict of the sovereign people, and was quietly driven away from the Élysée as Poher formed an interim government.

3

Echoes of the Occupation

'French people, you have short memories,' said Marshal Pétain in 1941.[1] It was, perhaps, not that they had short memories but that they refused to confront the truth about what happened in France under the German Occupation and wove a myth of the Resistance of the French nation that enabled them to absorb that past more easily into their conception of French history and French identity.

After the defeat of France in 1940, German forces occupied the northern half of France and set up a military administration to run it. In the southern part of France, which remained unoccupied, those who had long opposed the Republic came to power, set up an authoritarian regime at the spa town of Vichy under Marshal Pétain, abolished the Republic, and implemented a reactionary programme known as the National Revolution. The independence of the Vichy regime *vis-à-vis* the Third Reich was limited, and in any case it entered upon a path of 'collaboration' with Germany, in theory to lighten the burdens of the Occupation and to secure the return of French prisoners of war, in fact imitating many of the policies of the Third Reich and making it easier for the Germans to pursue their own policies. Thus Vichy not only passed legislation in 1940 and 1941 to exclude Jews from public, professional, and business life, but it co-operated with the German plan to deport Jews to Auschwitz and other extermination camps. Despite the immense popularity of Pétain, not all French people accepted the Vichy regime. Some found it too reactionary and gerontocratic and sought the patronage of the Germans in the hopes of forming a single Fascist party and pushing through a Fascist revolution. Others joined de Gaulle and the Free French in London or took part in the clandestine Resistance, especially after the Germans occupied the southern zone in 1942 and started to deport young men to work in German factories under the Compulsory Labour Service (STO) scheme of 1943. However, as Resistance grew, so repression increased at the hands of the Militia, which developed out of Pétain's Légion Française des Combattants (French Veterans' Legion) and eventually took over at Vichy, turning it into a police state. In the last months of the Vichy regime, after the Allied landings of June 1944, the French were engaged not only in fighting the Germans but in fighting each other, in a brutal civil war. The Liberation, finally, was accompanied by a purge of those in the Vichy regime, Militia, fascist movements, and others who were seen to have collaborated with or benefited from the German Occupation.

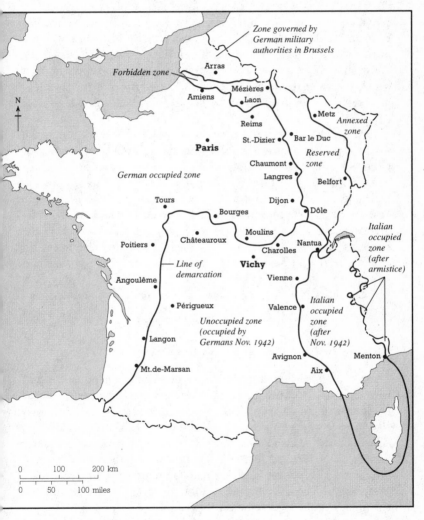

FIG. 3. Occupied and Unoccupied France, 1940–1942.
Source: W. D. Halls, *The Youth of Vichy France* (Oxford, Claredon Press, 1981).

The purge

There is no doubt that this purge was radical and sometimes bloody as collaborators, informers, and black marketeers were dealt with in an anarchic settling of differences by Resistance fighters and angry crowds. Figures of 100,000 summarily executed were bandied about in some quarters. Comparisons were made between the purge of the Liberation and the Terror of 1793. A much lower figure of 10,000 summary executions has generally been agreed by the ministry of the interior, by General de Gaulle in his memoirs, and by historians. On the other hand, the number of women who had their heads shaved for 'horizontal collaboration' in grotesque carnivalesque scenes by men who had been unable to protect either their country or their womenfolk has been raised by recent historians from a few hundred to a possible 10,000.

Such confusion about the numbers of casualties was not merely the result of poor statistical research. Mystification was a by-product of political rivalry. Those who were affected by the purges were keen to maximize the degree of their victimization in order to discredit the new regime that took power at the Liberation, while those who were determined that the Liberation should be a revolution rather than just the expulsion of the Germans were equally determined to make the purge as far-reaching as possible. Others, however, were intent on minimizing the purge in order to re-establish national reconciliation around the joy of Liberation, reconstruct French identity and French history in a way that restored French pride, and normalize political life.

Rivalry in the Liberation camp

There was, in fact, a struggle within the Liberation camp between hardliners and soft-pedallers that was almost as important as that between those who won and those who lost by the act of Liberation. The struggle was essentially between those who had participated in the Liberation as the clandestine movement of Resistance within France, and the Free French forces under de Gaulle who had set up a provisional government in Algiers and had returned to France in 1944 alongside the Allies. It reflected, to a large extent, a rivalry between the Communist and non-Communist Resistance.

For the former, the Liberation meant the settling of scores with those who aided, abetted, and benefited from the German Occupation and the Vichy regime at their expense. These were dealt with summarily, by revolutionary courts set up by the Committees of Liberation which replaced the disgraced or collapsing municipal and departmental authorities in the summer of 1944, or by courts martial set up by the integrated Resistance forces, the Forces Françaises de l'Intérieur (FFIs). Arguably the Communists acted with exaggerated ferocity because they had to purge the memory of the years 1939–41 when, as a result of the Nazi–Soviet pact, they had effectively been on the same side as Germany. Since the invasion of the Soviet Union by Germany they had borne the brunt of repression by the Germans and Vichy, and were now at the forefront of those for whom the Liberation meant the inauguration of a new, purified republic, staffed by a new political class forged by the Resistance and the sweeping away of discredited élites by administrative purges of the police, army, judiciary, civil service, business, and media. Intellectuals of the National Writers' Committee, for example, set up a committee to purge publishers who had worked for or been paid by the Germans, and drew up a blacklist of writers who were judged to have served the propaganda machine of the enemy.

To impose a revolutionary view of the Liberation, the National Resistance Council and Paris Liberation Committee, in which the Communists were highly influential, organized a civilian march from the place de la Concorde to the Bastille on 14 July 1945, complete with pikes and Phrygian bonnets, after the official military parade in the morning. The following month, they commemorated the Paris insurrection of August 1944 by unveiling a plaque at the entrance to the catacombs where the Communist Colonel Rol-Tanguy had had his headquarters and naming the place Stalingrad, in honour of the great Soviet victory over the German army, in the presence of the Soviet ambassador. Cynically, the Communist party also produced a forged document backdated to June 1940, calling for a popular insurrection and war of national liberation, in an attempt to construct a continuous pedigree of revolutionary-patriotic activity.

For de Gaulle, as head of the provisional government and the Free French forces, the Liberation had to be organized and presented in quite a different way. In the first place, the established channels of law and order and state power had to replace revolutionary authorities and revolutionary justice as soon as possible. *Commissaires de la République*, who were effectively prefects, were sent in to take over from Committees of Liberation and regular courts of justice were set up under a decree of June 1944 (though they were rarely active before October 1944) to deal with those facing charges. A High Court of Justice was also established to try ministers and high officials of the Vichy state, including Pétain and Laval. They were tried not according to revolutionary law but under article 75 of the penal code, which dealt with high treason. For those involved with Vichy in less serious ways, civic courts were set up alongside the courts of justice, with power to impose the punishment of national indignity, which deprived the condemned of the right to vote and banned them from all elective, government, or military office, and from many professions. This saved some associated with Vichy from harsher penalties imposed in regular courts, but had the effect of destroying their public careers. It was compounded by categories of ineligibility for the new elective assemblies, which included the 569 deputies and senators of the Third Republic who had voted full powers to Marshal Pétain on 10 July 1940 (only 80 had voted against).

The effect of these exclusions was to eliminate most of the old political class from political life in the immediate aftermath of 1945. Of those elected to the first Constituent Assembly, 85 per cent had never sat in parliament before the war and over 80 per cent had been active in the Resistance. This clearly carried risk for de Gaulle at the head of the provisional government. It threatened to impose the hegemony of parties of the Left, in particular the Communists, while removing traditional parties of the Right from the political scene. Moreover, though de Gaulle's own legitimacy was based on the fact that he had gone on fighting the war in 1940 while Pétain had capitulated and collaborated, to insist for too long on the division between 'good' resisters and 'bad' collaborators undermined the work of national reconciliation that he was seeking to undertake. Thus, having spoken during the war on the need for a great wave to sweep away the discredited French ruling class, at the Liberation de Gaulle declared that 'apart from a handful of wretches the vast majority of us were and are Frenchmen of good faith.'[2] This effectively drew a veil over the past, brushed away the civil war, and established the orthodoxy that virtually all French people had either participated in or sympathized with the Resistance. It also made possible much greater continuity between the Vichy regime and the Fourth Republic in the administrative, profes-sional, and business élites than among politicians. Purging commissions were basically internal operations run by the élites themselves, and *esprit de corps* offered protection for colleagues whose conduct had been wanting. Competence and the difficulty of finding replacements told more than political correctness. Of a million civil servants, just over 11,000 received some sanction (although recent research suggests be-tween 22,000 and 28,000) and only 5,000 were removed from office.

Some businesses were nationalized, more because they had collaborated than for economic reasons, but the task of economic reconstruction required a close partnership between government and business, not persecution. Even less serious was the purge of show business, for the sake of keeping up morale. Popular singers such as Fernandel, Arletty, Chevalier, and Piaf received no punishment for entertaining German audiences in occupied Paris; Mistinguett was let off with a reprimand.

To restore the continuity of French history was as important as to restore the unity of the nation. This could be achieved by presenting the wartime experience in a way best calculated to flatter de Gaulle, the army, and the nation as a whole. To this end, commemoration of the war began promptly in 1945. The anniversary of de Gaulle's appeal from London, on 18 June 1945, was celebrated by a show of force by the Free French, including tanks and planes, eclipsing the khaki of the FFIs, to privilege the contribution of regular soldiers at the expense of that of the clandestine Resistance. The military parade of 14 July 1945 was designed to mark the reunion of army and nation, and to outshine the revolutionary march organised by the Communists in the afternoon. The armistice commemoration of 11 November 1945 underlined de Gaulle's message that the defeat of Germany in 1945 marked the end of a Thirty Years War, to marginalize the defeat of 1940 and the Occupation as blips in French history, and to equate the victory of 1945 with that of 1918. Descending the Champs-Elysées to lay a wreath at the statue of Clemenceau and a palm at that of Foch, de Gaulle sought to demonstrate a continuity between the victorious leaders of 1918 and himself as leader of the Free French, and to suggest that the Republic had never died between 1940 and 1944. In a much longer historical perspective 8 May, date of the German surrender in 1945, celebrated for the first time as a public holiday in 1946, was honoured by a visit by de Gaulle to the grave of Clemenceau in the Vendée, and happily coincided with the festival of Joan of Arc, who had expelled the invader from France so many centuries before.

Return of the ghosts of Vichy

Before he was shot, Joseph Darnand, the head of the Militia, remarked that legitimacy had changed sides at the Liberation. This was a shrewd judgement. After the Liberation, activity in the Resistance (or successfully laying claim to it) was the prerequisite of holding political office in the Fourth Republic. The successful imposition of the 'Liberationist' version of history by the architects of the new Republic made it very difficult for those tainted by association with Vichy to make a come-back.

Gradually, however, those ghosts began to reappear. A right-wing press started to function, a much-publicized banquet was organized in March 1948 by the Association of Representatives of the People of the Third Republic for 1,000 deputies and mayors who had been dis-qualified from political life, and the following month Jacques Isorni, who had defended Marshal Pétain at his trial, set up a committee to obtain his release from prison. A counter-orthodoxy to that of the Liberation was developed around homage to Marshal Pétain. It argued, first, that the armistice was not capitulation but the right decision, beneficial to the French as a whole; second, that the Vichy regime had been the legitimate government of France, established by parliament in July 1940, so that those who served it were executing legitimate orders and should not be punished; third, that the purges were worse than the Terror of 1793 in terms of the numbers killed and the scandal of revolutionary justice; and fourth, that a sweeping amnesty of those condemned at the Liberation was thus a matter of urgency.

The question of an amnesty could not be avoided by the politicians of the Fourth Republic. De Gaulle, out of power, conceded in 1949 that Pétain, who symbolised capitulation and collaboration, had been right-ly condemned, but that it was unfair to leave an old man of 93 locked up in a fort on the Atlantic Île d'Yeu. Vincent Auriol, the Socialist president of the Republic, accepted that some justice at the Liberation had been harsh, and declared himself ready to accept an amnesty except for a hard core of traitors, informers, torturers, and convicted collabor-ators. A poll in the *Figaro* in June 1949 showed 60 per cent in favour of an amnesty, 23 per cent against.[3] The Gaullist RPF and the MRP were in favour of an amnesty on the grounds of natural justice, clemency, and the need to restore national unity, while Communists and Socialists opposed it, fearing that to grant amnesty to guilty men would rehabilitate and excuse them. But Communists, who had tacitly been held responsible for many of the 'crimes' of the Resistance, had been excluded from the governing coalition in 1947 as the Cold War began to bite. The moment was therefore propitious for an amnesty which would forge a new kind of national unity, exclude the Commun-ists, and restore the Right to the mainstream of political life.

A first, limited amnesty was voted in January 1951. Among other things, it lifted the sanction of national indignity from the heads of many politicians and returned them to the political nation. The general election of June 1951 saw the return of significant numbers of right-wing deputies for the first time since 1945. One of them, Jacques Isorni, was elected in the posh sixteenth *arrondissement* of Paris with the help of nationalist students led by Jean-Marie Le Pen. Isorni responded to the death of Marshal Pétain in July 1951 by setting up an Association to Defend the Memory of Marshal Pétain, which campaigned to revise the verdict of his trial, to have his body translated from the Île d'Yeu to Verdun, scene of his triumph in 1916, and to rehabilitate the values of Pétainism. He also led a campaign for a second, more general amnesty, citing de Gaulle's view that only 'a handful of traitors' were guilty, arguing 'that there had not really been two blocs, the good and the bad, heroes and traitors; no more had Vichy itself been a bloc.'⁴ A second amnesty became law in August 1953, and the number of prisoners held, which had been 4,000 in January 1951 and 1,570 in October 1952, fell to 62 in 1956 and 19 in 1958; all were out by 1964.

The granting of the amnesties threatened to upset the difficult balance between the legitimacy established by the Resistance as the touchstone of holding office in the Fourth Republic and the need for national reconciliation and unity. To privilege the latter ran the risk of calling into question the former. In February 1953, while the second bill was being debated, the question arose again when 21 members of the SS division *Das Reich* were put on trial before a military court at Bordeaux. It was alleged that on 10 June 1944, as they headed north to counter the Allied invasion, they had massacred the inhabitants of the Limousin village of Oradour-sur-Glane by imprisoning them in the church and setting it alight. The complication was that twelve of the accused were not Germans but from Alsace, which had been re-annexed by Germany in 1940, and had been conscripted by force into the Waffen SS. Their claim was that they had fought for the Germans, but against their will, 'malgré nous', out of fear of reprisals against their families if they did not. Most of the Alsatians were nevertheless sentenced to terms of forced labour, provoking an outcry in Alsace. The government feared the revival of the autonomist movement that had flourished in Alsace between the wars, in opposition to the Third Republic's policies of anticlericalism and Frenchification, and proposed an amnesty to the National Assembly. But concessions to maintain national unity in turn challenged the issue of the illegitimacy of the Occupation, and angry deputies from the Limousin replied that an amnesty to mollify the Alsatians insulted the families of victims of the massacre. In the end the amnesty was voted, provoking strikes in the Limousin and pilgrimages to the martyred village. The punishment of wartime atrocities had to give way to the demands of national unity.

The myth of Resistance

Forced onto the defensive, those who believed that legitimacy in the Republic was determined by participation in the Resistance and subscription to its values began to mobilize in order to put their view across. In 1951 a Committee for the History of the Second World War was established, attached to the prime minister's office. Its president was the eminent historian Lucien Febvre, but the real force behind it was Henri Michel, who had been active in the Resistance and was determined to impose both an academic orthodoxy of the Resistance and the Resistance as a civil religion. The following year, a National Association of Ex-Servicemen of the French Resistance (ANACR) was set up. Its aim, in the light of the amnesty for 'war criminals', was to obtain the release of the 100 Resistance fighters still in prison for 'crimes of Resistance', to combat the right-wing view that the Resistance were no better than bandits or terrorists, and to obtain the status and privileges of ex-servicemen for those who had fought in the Resistance. In the orbit of the ANACR, hundreds of local Resistance associations were set up, each seeking to set the record straight about small Maquis organizations and actions and to sustain the camaraderie of those who had fought in them. The coming to power of de Gaulle in 1958 gave a boost to these movements. A *Figaro* poll of 1970 revealed that 52 per cent of French people still saw him as 'the man of 18 June 1940', 32 per cent as 'the man of the Liberation', and only 12 per cent as 'the founder of the Fifth Republic.'[5] De Gaulle endorsed the cult of the Resistance while underlining his own Resistance credentials in 1964 by presiding over a ceremony transferring to the Panthéon the remains of his trusted lieutenant, Jean Moulin, who had served as link man between de Gaulle and the internal Resistance and died at the hands of the Gestapo.

The historical orthodoxy elaborated by Resistance historians, Resistance organisations, and de Gaulle rested on four key articles of faith. The first was that the Resistance had been a heroic struggle, with a long roll-call of martyrs, with at least 30,000 shot (the Communists claimed that 75,000 Communists alone had been shot) and 115,000 deported, of whom only 40,000 returned. This was designed to counter accusations of the banditry and 'crimes' of the Resistance and to deprive partisans of Vichy of a monopoly of victims. The second was that the Resistance had a pedigree going back further than the Vichy regime itself, to the appeal of Charles de Gaulle on 18 June 1940 that established his legitimacy and leadership of the Resistance. The third was that there was a coherent ideology of the Resistance, centring on the defence of the rights of man, and that all who participated, no matter how varied their backgrounds and motivations, subscribed to these values. The final point was that, though active resisters were a minority, they had been able to operate because of the support of the nation. The Resistance recreated national unity, and in turn imparted the grace of having participated in the Resistance to the nation as a whole.

The mirror cracked

This redeeming, unifying, heroic story of the wartime years, carefully elaborated in the 1950s and 1960s, was shattered in the 1970s. Much of the responsibility for this lay with a film of Marcel Ophuls, *Le Chagrin et la Pitié* (*The Sorrow and the Pity*), which opened in Paris in April 1971 but which was not screened on television until 1981. Ophuls had fled to France from Germany in 1933, then fled France for the United States in 1941, returning to France in 1950. His film, subtitled 'Chronicle of a French town under the Occupation', interviewed the ordinary people of Clermont-Ferrand as well as the Wehrmacht officer who had been in charge of it. The film exposed not the heroism of the French but their guilt, not their unity but their divisions. They were revealed to have been passive, mediocre, fearful, hateful, deceitful, and self-deceiving. Their instinctive Pétainism was highlighted, together with their anti-Semitism. For the first time the anti-Jewish policies of the Vichy government and their acceptance by the vast majority were made explicit.

Some responsibility for this revision of ideas also lay with France's leaders. While de Gaulle sought to embody and interpret the Resistance, his successor as president, Georges Pompidou, had taught at the Lycée Henri IV all the way through the Occupation and had reprimanded a pupil for taking down a portrait of Marshal Pétain that hung in the classroom. In 1971 he told the *New York Times Magazine* of the 'irritation and loathing which the Resistance inspired in him.'[6] Pressure was put upon him from clerical circles to offer a pardon to the head of intelligence of the Militia in the Lyon region, Paul Touvier, who had twice been condemned to death, in 1946 and again in 1947, for various atrocities and had since been sheltered by the Church. De Gaulle had said, 'Touvier? the firing squad for him', but Pompidou pardoned him secretly in November 1971. When the news was scooped by *L'Express* and an outcry began, Pompidou asked the press: 'Has not the time come to throw a veil, to forget those times when the French did not like each other, tore each other apart, even killed each other?"[7] As it happened, the corner of the veil was only just beginning to be lifted.

It has been argued that Pompidou was fishing for the support of the right-wing Independent Republican party of Giscard d'Estaing, which he needed for the elections of 1973. Giscard, elected president on Pompidou's death, was the grandson of a member of Vichy's National Council, and had no background in the Resistance although he had joined the First French Army for the invasion of Germany in December 1944. In 1975 he demoted celebration of victory in Europe, 8 May, from the status of public holiday, ostensibly in the name of Franco-German *rapprochement* and because neither Britain nor the United States marked it, but also out of deference to the Right. This gesture was compounded by his sending of the prefect of the Vendée to lay a wreath at the tomb of Marshal Pétain on the Île d'Yeu on Armistice Day 1978, defending his decision by pointing out that Pétain was a marshal of two world wars.

In the presidential election of 1981 François Mitterrand was presented as the candidate of the Resistance, pitted against Giscard, from a family of collaborators. Giscard's supporters pointed out that Mitterrand had worked for prisoners of war under the auspices of Vichy, and had been decorated with its *francisque* medal before going over to the Resistance. They were also keen to assert that Mitterrand's Communist ally, Georges Marchais, far from being deported as an STO worker, had gone voluntarily to Germany to work in the Messerschmitt factory. Mitterrand's career was indeed typical of that of many others, who had not resisted from the first hour on 18 June 1940 but had gone over at a later date. As president, moreover, Mitterrand came to behave less as a Socialist president and more as the president of all the French people. In accordance with this, he had a wreath laid on the tomb of Pétain in September 1984, when he shook hands with Chancellor Kohl at Verdun, again in June 1986, to mark the 70th anniversary of Verdun, and every armistice day after 1987. In the summer of 1992 Mitterrand was urged by Jewish organizations to stop honouring the grave of Pétain, whom they held responsible for deporting 75,000 Jews. The prefect of the Vendée still travelled to the Île d'Yeu on 11 November 1992, but the following year Mitterrand agreed to lay the wreath at Verdun, to highlight Pétain's role in the First World War, not in the Second.

The Holocaust factor

While the desire for national reconciliation suggested to many that at worst the balance of good and evil should be allocated fairly between the Resistance and Vichy and at best the period should be forgotten altogether, to others it was a matter of the greatest urgency that the truth about the past as they knew it should first be told and secondly become the accepted historical truth.

One group that had been entirely marginalized in accounts of the Occupation and now sought redress were the Jews. The National Movement of Prisoners or War and Deportees (MNPGD), established in March 1944 by François Mitterrand and others, sought to bring together prisoners of war and STO workers as well as those deported to Nazi concentration camps. In 1945, after the POWs and STO workers founded separate organizations, the organization of deportees was effectively left to the Communist party. The Communists, however, were interested only in political deportees, who included many Communists, and in obtaining for them the status and privileges of Resistants. Jewish survivors of concentration camps were of no concern to them. The Communists were also keen to demonstate that Auschwitz was essentially a camp for political prisoners, not least Communists, rather than a camp for exterminating Jews. The Gaullist account of the deportation also had the effect of marginalizing Jews. In 1962 a crypt on the Île de la Cité in Paris was dedicated to the 200,000 'martyrs of deportation' to Nazi camps. Since most of the Jews deported from France had not in fact been French nationals, it is unclear whether they were explicitly remembered there. Moreover, the image of Jews in France was not improved by de Gaulle's declaration after the Six-Day War in 1967 that 'the Jews . . . who had always been an élite people, self-confident and domineering, would, once gathered in the land of their former greatness, transform the very moving desires they had formed for nineteen centuries into a burning and conquering ambition.'[8]

Against this deadweight of prejudice and silence, those who spoke for the Jewish community campaigned to establish as historical fact the French dimension of the Holocaust, when both French and non-French Jews had been deported to death camps, the Nazi occupiers being greatly helped in their task by the French authorities in Paris and Vichy. It was not until the 1970s that the message about Vichy and the persecution of the Jews began to be established. The central event in the account was the rounding up of 12,000 mainly non-French Jews on the night of 16–17 July 1942, to be crammed into the Vélodrome d'Hiver sports stadium without food or water for several days before transportation via Drancy to Auschwitz. In the first major study of this event, published in 1967, the authors complained that they had been denied access to the relevant archives by the French government, but they demonstrated that the French police had made available a file naming 27,388 Jews in Paris, and that the French police and fascists had rounded up the Jews in an operation organized in conjunction with the Commissariat for Jewish Questions under Darquier de Pellepoix and the Gestapo in Paris, in accordance with orders from Reinhard Heydrich. After 1971 Serge Klarsfeld, whose father had been deported to Auschwitz, and his wife, Beate, ran a double operation, first to track down Nazi war criminals and bring them to justice and secondly to demonstrate the responsibility of the French state for the persecution and deportation of Jews. When President Giscard d'Estaing visited Auschwitz in 1975 and declared that 110 French people, including 48 Jews, had been deported there, Klarsfeld replied by publishing details of 70,000 Jews who had been deported from France to Auschwitz, 23,000 of them French and 47,000 foreign, while 1,000 French and 4,000 foreign Jews had been deported to other camps.[9]

That this view of the Holocaust and the French involvement in it would gain general acceptance was in no sense a foregone conclusion. Darquier de Pellepoix, who had been condemned to death *in absentia* at the Liberation and taken refuge in Spain, gave an interview to *L'Express* in October 1978 in which, while revealing the role of the Vichy police chief René Bousquet in the round-up of the Vel d'Hiv, asserted that the only things gassed at Auschwitz had been lice. The following month Robert Faurisson, a lecturer in French literature at Lyon university, had published in *Le Monde* at the twenty-third attempt an article entitled 'The Problem of the Gas Chambers and the Rumour of Auschwitz'. Here he argued that the gas chambers had never existed, that genocide had never happened, and that contrary assertions were nothing but Zionist lies.[10] This provoked demonstrations at Lyon and the suspension of Faurisson's lectures, but he found a ready audience at a congress of revisionists in Los Angeles in 1979 and through the *Journal of Historical Review*, launched in 1980. The strength of anti-Jewish feeling was exemplified by a bomb explosion outside a synagogue in the rue Copernic, Paris, on 3 October 1980, which killed four and wounded twenty.

Needless to say, the so-called revisionist arguments were squarely attacked. Some took direct action, burning books at the revisionist bookshop and publisher, La Vieille Taupe, in February 1981. The academic community closed ranks against Faurisson. The ancient historian Pierre Vidal-Naquet, whose mother had died at Auschwitz, denounced Faurisson as 'a paper Eichmann', and used the critical methods of the historian to expose the sham of the revisionist case. A chair in the history of the Shoah was founded at the Sorbonne, giving official sanction to the fight against revisionism. For their part, Jewish leaders managed to establish the anniversary of the Roundup of the Vel d'Hiv as a central act of commemoration on its 50th anniversary, 16 July 1992, when François Mitterrand agreed to attend the monument on the site of the stadium. It was an occasion of mixed sentiments, since Mitterrand had refused officially to accept the responsibility of the French state for crimes committed against Jews, as the Republic was not Vichy and Vichy not the Republic. He was thus both booed and applauded by the demonstrators. The last word was had by Robert Badinter, the minister of justice and himself Jewish, who agreed that the Republic did owe an ultimate homage to the victims of Vichy, 'the teaching of the truth and the force of justice.'[11]

The revisionist arguments, though unacceptable in public discourse, nevertheless did have a resonance, at least on the extreme Right. Interviewed in 1987, Jean-Marie Le Pen stated, 'I don't say that the gas chambers did not exist. I haven't been able to see any myself. I have not made a special study of the question. But I think that they are a point of detail in the Second World War.'[12] A poll in 1990 showed that while only 1 per cent of those aged 18–44 thought that the existence of the gas chambers was a lie and 65 per cent thought their existence clearly proven, 33 per cent, however, believed that they had existed although the fact had not been adequately proven.[13] Appearing in court in 1991 for claiming that the myth of the gas chambers was 'scandalous', and fined, Robert Faurisson was nevertheless able to denounce 'sex-shop anti-Nazism' and the Holocaust religion, and likened himself to a victim of the Inquisition burned at the stake because he did not believe in the devil.[14]

'What is left of the Resistance?'

The Holocaust orthodoxy clearly established the Jewish people as the victims of persecution and genocide. But while their identity as victims legitimated their cause, it also conferred on them the image of a people who had gone like lambs to the slaughter. Another, heroic identity of the Jews as Resistance fighters was required to balance that of passive martyrdom.

The Jewish dimension of the Resistance was effectively ignored until 1985 when a film, *Terrorists in Retreat*, shown on television, highlighted the activities of the group led by the Armenian poet Missak Manouchian. While the film dealt with the role of immigrants in general in the Resistance, Klarsfeld's Association of Sons and Daughters of Jewish

Deportees from France was quick to claim credit for the Jewish contribution. It had been axiomatic in the official account elaborated by Henri Michel and his school that the Resistance had been a bloc, fighting for the same humanitarian, republican French values, converging on the Liberation, the Fourth Republic, and the recovery of French national greatness; there was no room for an interpretation that the Resistance had been a melting-pot of freedom fighters, each struggling for a different cause. During the Occupation itself, the Germans had tried to discredit the Resistance as a movement of immigrant terrorists and international agents. In February 1944 they organized a show trial of twenty-three such Resistants, including five Italians, two Armenians (Missak and Armenek Manouchian), a Spaniard, a Pole, and twelve Jews, seven of whom had been born in Poland, three in Hungary, one in Bessarabia and one in France of immigrant Polish parents. The twenty-two men were immediately executed by firing squad, the one woman taken to Germany and later guillotined. It was all the more important, therefore, at the Liberation, to demonstrate the unity and Frenchness of the Resistance. Thus in the march-past in liberated Toulouse, on the orders of the *Commissaire de la République*, the Maquis of the Jewish Combat Organization (OJC) were not allowed to carry the blue and white Jewish flag, and the Spaniards were not allowed to carry that of the Spanish Resistance.

The division between assimilated French Jews and non-assimilated immigrant Jews did not help matters. While the former were part of the French middle class and happy to fight for the liberation of France to underline their own assimilation, those who had immigrated from Russia and eastern Europe after 1917 spoke Yiddish and, together with Jewish refugees from central Europe after 1933, were effectively organized by the Communists in their Main d'Œuvre Immigrée (Immigrant Labour, or MOI) organization. This was divided into German, Czech, Hungarian, Romanian, and Yiddish (mainly Polish) language groups, many if not most of the non-Yiddish groups also being Jews. When the Communists entered the struggle against Hitler, they organized the MOI into branches of their guerilla movement, the Franc-Tireurs et Partisans (FTP), so that many of the most active FTP were immigrant Jews, fighting under Communist colours. After the war, however, the MOI was dissolved and these immigrants were sent back to eastern Europe by the Communists to build socialism there; many did not return to France until the 1960s. Not until after the round-up of July 1942 and the Warsaw Ghetto uprising did Jews set up the Jewish Union for Resistance and Mutual Aid and the OJC, to fight less for the

French Resistance or for Communism than for the survival of Jewry. The story of the Jewish contribution to the Jewish Resistance took even longer to come out than the Jewish contribution of the French Resistance. But together they had the effect of shattering the myth of the unity and Frenchness of the Resistance.

Not only the unity but also the heroism of the Resistance came under question. The official orthodoxy of Henri Michel and his school, which privileged the leadership of de Gaulle from 18 June 1940 and the co-ordination of the internal Resistance by his *alter ego*, Jean Moulin, had long been contested by Communist participants and historians who gave priority to the internal Resistance and national insurrection. Neither side called into question the heroism of the other, only its primacy. Later, however, controversy broke out which questioned the fundamental integrity of leading figures of the Resistance. Henri Frenay, founder of *Combat*, an early Resistance organization in the unoccupied zone, insinuated that Jean Moulin had secretly been working on behalf of the Communists. In 1989, a year after Frenay's death, Moulin's former secretary, Daniel Cordier, replied with the first volumes of a multi-volume study of Moulin. Posing as objective history, Cordier settled his scores with Frenay by attributing to him a manifesto dated November 1940, which showed that Frenay had been Pétainist, a supporter of Vichy's National Revolution, anti-Semitic, and had accepted the official strategy of collaboration with Germany if only as a ruse behind which a 'movement of national liberation' could organize. This publication was a bombshell, provoking attacks on Cordier by Frenay's family, former members of *Combat*, and defenders of the Resistance orthodoxy. The overall effect was of the Resistance washing its dirty linen in public and destroying the heroic interpretation that had held sway for so long. Moreover, once the dam had been breached, those who wished to discredit the Resistance as a whole were allowed free play. In 1993, for example, the journalist and expert on espionage Thierry Wolton denounced Jean Moulin as a Soviet agent, a view that would have been ridiculed had it not been given the seal of serious research by respectable historians (and ex-Communists) Annie Kriegel and François Furet. After twenty years of attacks on the Vichy regime, the guns now turned back on the Resistance. As the review *Esprit* put it in 1994, 'What is left of the Resistance?'[15]

As the Resistance lost its identity of heroism and unity it opted for another strategy: to throw in its lot with the victims of the Holocaust. Traditionally, the heroes had not wanted to mix with the victims; the active clandestine movement had been concerned to distinguish itself from the passive objects of deportation. The capture and forthcoming trial of Klaus Barbie, the Gestapo chief in Lyon who was alleged to have killed Jean Moulin as well as a number of Jews, caused them to change their minds. Under a French law of 1964, while war crimes could no longer be tried twenty years after their perpetration, crimes against humanity became imprescriptible and could be tried at any time. Barbie was charged with crimes against humanity, and in order to win compensation Resistance organizations sought a ruling that crimes against Resisters as well as crimes against Jews were also crimes against humanity. In December 1985 the Supreme Court of Appeal obliged by redefining crimes against humanity as

inhuman acts and persecutions committed systematically in the name of a state practising a policy of ideological hegemony against persons by virtue of their belonging to a racial or religious group, but also against the opponents of that policy, whatever the form of their opposition.

Thus, ironically, while apologists of the Jewish community had campaigned to secure a status as Resisters, not just victims of persecution, apologists of the Resistance secured a decision giving them the same status as persecuted Jews.

Barbie and Touvier: the trivialization of evil

For those who defended the memory of the Jewish experience the Holocaust was a unique event, which defied comparison with any persecution or extermination before or since. Simone Veil, former minister of health who had herself been deported to Auschwitz, was scandalized by the eradication of the difference between Resisters and Jews which trivialized the concept of the crime against humanity and relativized the Holocaust. But this was precisely the agenda of those who sought to exculpate those accused of crimes against humanity. For them, it was crucial to demonstrate that the persecution, even extermination, of the Jews was no worse than any other atrocity in history, some of which had been perpetrated by the French state.

Klaus Barbie, the head of the Gestapo in Lyon, who had been condemned to death *in absentia* in 1952 and 1954, gone into hiding in Bolivia, been tracked down in 1971, and finally brought back to France, went on trial before the assize court of Lyon between May and July 1987. Serge Klarsfeld, who had long pursued him and was himself a lawyer, provided most of the case against Barbie in the court. He concentrated on the round-up of 44 Jewish children from a Red Cross colony at Izieu, not far from Lyon, who were deported to Auschwitz between April and June 1944, and of whom only two survived, and on the organization of the last convoy from Lyon on 11 August 1944. But in Barbie's defence team, led by Maître Vergès, Klarsfeld had formidable opponents. Vergès fought to turn the tables on the prosecution, by arguing first that the Resistance was riddled by informers who were as responsible for the fate of Jean Moulin as Barbie was, and secondly that the atrocities perpetrated by the Nazis against the Jews were no worse than those committed by the French in their colonies. Vergès himself had a Vietnamese mother, and his wife was an Algerian freedom fighter who had been tortured by the French during the Algerian war; the other two defence lawyers were a Congolese, Maître M'Bemba, and an Algerian, Maître Bouaïta. It began to appear that France rather than Barbie was on trial. Vergès pointed out that, on the day of victory in Europe, 8 May 1945, the French had massacred 15,000 Algerians at Sétif. 'How many Oradours can you get into that?' he scorned. Maître M'Bemba argued that the building of the Trans-ocean railway from Pointe Noire to Brazzaville in French West Africa, which had cost 8,000 African lives, was worse than the last convoy out of Lyon and had probably given Hitler ideas about the organization of genocide. Maître Bouaïta, for his part, said there was no difference between a crematorium furnace and a phosphorus bomb, nor between the atrocities committed by the Nazis and those perpetrated by the Americans in Vietnam or the Israelis in the Palestinian camps of Sabra and Chatila in Lebanon in 1982. Though Pierre Vidal-Naquet replied that brutal French colonialists had acted against the laws of the Republic while Himmler and Eichmann had acted in accordance with Hitler's principles, the philosopher Alain Finkielkraut had to admit that Vergès had produced a certain effect by 'hammering home that Auschwitz was not the anus of the world but the navel of the West.'[16]

For all the pyrotechnics of Vergès, Barbie was found guilty of crimes against humanity and sentenced to life imprisonment. This caused the French little grief, since Barbie was a German and a Nazi. But in May 1989 the Militia leader Paul Touvier was finally caught and arrested. There began a long process of buck-passing by the courts until in April 1992 the court of criminal appeal of the appeal court of Paris dismissed the case, on the grounds that insufficient evidence existed for five of the six crimes of which he was accused, while the sixth, the massacre of seven Jewish hostages at Rillieux-la-Pape near Lyon on 28 June 1944, in reprisal for the assassination of the Vichy information and propaganda minister Philippe Henriot, was not a crime against humanity. The reason given was the technical one that

this crime was not committed in the execution of a concerted plan carried out in the name of a state practising a policy of ideological hegemony and to achieve the extermination of civil populations or any other inhuman act, or persecution for political, racial or religious motives.

The explanation was that, while the Militia may have had hegemonic ambitions, the Vichy state did not. Indeed it had no 'precise ideology', only 'a constellation of "good sentiments" and political animosities', mainly directed against Communism. Further, Vichy never officially proclaimed, as did the Third Reich, that the Jew was an enemy of the state, while Marshal Pétain never made any anti-Semitic comments in his speeches.

The 215 pages of this ruling were a thunderclap. For Touvier's lawyer, whose father had defended collaborators at the Liberation, and for the extreme Right, it was a vindication of Vichy and a legitimation of their views. François Brigneau, a former Militiaman and editor of the extreme-right *National Hebdo*, said that the ruling 'honoured the French judicial system as a whole' and ended 'half a century of civil war.'[17] Most of the political class, many in the Church who had formerly protected Touvier, and 73 per cent of French people polled by *Le Parisien* were shocked by the ruling. It threw doubt on the soundness of the judicial system and illustrated the force of collective amnesia. The Gaullist Patrick Devedjian hoped that 'our country would grow by coming to terms with its past. Liberation will come only by admission and forgiveness, not by denial.'[18] There were demonstrations in Lyon, Grenoble, and Chambéry, the old haunts of Touvier, and a march to the Memorial of the Deportation in Paris. The film director Claude Chabrol was so disgusted by the whitewashing of Vichy that he decided to make a film composed only of the regime's propaganda, *The Eye of Vichy*.

Under the weight of protest, the Supreme Court of Appeal decided in November 1992 partially to overrule the appeal court of Paris, stating that Touvier could be sent to trial for the Rillieux-la-Pape massacre. On 8 June 1993, before the trial opened, René Bousquet, the Vichy police chief whose role in the round-up of the Vel d'Hiv had been exposed by the interview of Darquier de Pellepoix in 1978 and more recently in the Chabrol film, was assassinated at his Paris flat. While the general view was that a lone fanatic was responsible, it was also argued that the murder suited right-wing groups who did not want Vichy's role in the Holocaust further exposed. Whether Vichy would have emerged more damaged from such a trial is difficult to gauge; certainly, when the trial of Paul Touvier took place at the Assize court of Versailles in March–April 1994 there was an outside chance of an acquittal.

Touvier's defence counsel, Maître Trémolet de Villers, did not contest the fact of the massacre of Rillieux-la-Pape. He argued that it was a 'terrible but minor act of war' in reprisal for the assassination of Henriot, and that Touvier had reduced the number of Jews executed to seven while his superior had proposed thirty and the German Gestapo at Lyon were talking of a hundred. That said, Mc Trémolet argued that the massacre was a French act, and had nothing to do with German orders. The prosecution, by contrast, argued that 'the plan was Nazi, the complicity French.'[19] The concerted plan for the systematic extermination of the Jews for the benefit of a state practising ideological hegemony was pinned not on Vichy but Nazi Germany. What was then required was to demonstrate Touvier's knowledge of that plan and complicity in it.

Touvier was found guilty of crimes against humanity, the first Frenchman to be thus condemned, and sentenced to life imprisonment. The following weekend, Jewish students demonstrated outside the Paris home of Maurice Papon, who had been accused of crimes against humanity in 1983 and 1992. Papon had been secretary-general of the prefecture of Bordeaux between 1942 and 1944, when he was responsible for the deportation of 1,700 Jews, prefect of police in 1961, when he was responsible for the deaths of 200 Algerians demonstrating against the Algerian war, and Raymond Barre's budget minister in 1978 before his past was revealed by the satirical magazine *Le Canard enchaîné* in 1981. Gérard Boulanger, a barrister representing families of deported Jews seeking compensation from Papon, was so disgusted by Papon's assertion in 1990 that he was 'the Dreyfus of modern times' that in 1994 he published a book entitled *Maurice Papon, a French Technocrat Under the Occupation*, to expose the 'amoral ambition of a cold technocrat'.[20] Papon, incensed, brought libel proceedings against Boulanger, but was unable to secure a conviction from the Tribunal Correctionnel of Bordeaux in June 1994. The appearance of the 83-year-old Papon before a criminal court, meanwhile, was still awaited.

The Mitterrand affair

It would have been soothing if in 1994, the 50th anniversary of the Liberation, some degree of national reconciliation had been achieved around the joy of Liberation, while recognizing objectively that life under the Occupation had been extremely difficult and that choices were made that had sometimes been cause for regret. Instead, the nation was divided and confused by revelations about the role of François Mitterrand in a book by Pierre Péan, *Une Jeunesse française*. This book was all the more controversial in that it was based on interviews with Mitterrand and in some sense authorized by him.

Naturally, a certain amount about Mitterrand's record and attitudes was already known: his work for prisoners of war under the Vichy government, the award of the *francisque*, his insistence on laying a wreath at the tomb of Marshal Pétain, and his refusal to take full responsibility for the persecution of the Jews on behalf of the French state. After the Touvier trial he was reported to have said, 'You cannot live the whole time on memories and grudges,' and that the persecution of Jews was part of the logic of the Second World War.[21] It was now demonstrated, however, that he had joined the Volontaires Nationaux, the section of the extreme right-wing Croix de Feu open to those too young to have fought in the First World War, as soon as he arrived as a student in Paris in 1934; that he had taken part in a demonstration to protest against 'the invasion of immigrants' in 1935; that he had been a devoted Pétainist, a member of the Légion des Combattants which spearheaded the National Revolution, had been put in charge of organizing repatriated POWs for the benefit of that Revolution, and had been introduced to the Marshal himself; that he had gone over to the Resistance in 1943 but had no liking for de Gaulle and acted more out of hostility to the Germans than hostility to Vichy; that he had been a friend of Vichy police chief René Bousquet, whom he met in 1949, after Bousquet had been acquitted by the High Court of Justice on charges of collaboration just as Mitterrand, as secretary of state for information, had announced the first bill to grant amnesty to collaborators, and that each had subsequently helped to further the career of the other.

In some ways, there was nothing uncommon about much of this. 'Nonconformists of the 1930s' were as likely to be found on the extreme Right as on the extreme Left, and switches between the extreme Left and extreme Right and vice versa were frequent between 1935 and 1945. The transition from loyalty to Vichy to loyalty to the Resistance or Free French was common, even among Vichy officials. The myth of the Resistance as a bloc and a clean break with the past had long been exploded: there was more continuity between the political class of Vichy and that of the Fourth and Fifth Republics than had previously been admitted. Given that the decisive factors in Mitterrand's political career were opposition to Communism on the one hand and opposition to Gaullism on the other, his somewhat cerebral and opportunistic espousal of socialism in later life became understandable.

What was different about the Mitterrand affair, however, was that his choices in the 1930s and under Vichy, and his enduring Pétainist sympathies and friendships, called into question the sincerity of his embrace of socialism and the Left. He had claimed the mantle of the Popular Front, but had opposed it in 1936; he had attacked Giscard d'Estaing in 1981 for his links with Vichy when his own were far more explicit; he had cast himself as the successor to Jean Jaurès while continuing to see René Bousquet until 1986. More serious, however, was the fact that he had won over the loyalty and devotion of generations of socialists and partisans of the Left, who believed him to be the right man to head the new Socialist party, then ensure the endurance of the Left in power, while all the time he had been hiding a past as little better than a collaborator. It was no accident that while politicians of the Right were indulgent towards or embarrassed by Mitterrand's revelations, perhaps because of the complexity of the relationship of the Right with Vichy, those on the Left were disorientated and struck by a sense of betrayal. Pierre Mauroy, who had supported Mitterrand's bid to become leader of the Socialist party in 1971, and had been his first prime minister in 1981, tried to draw a distinction between Mitterrandism and socialism, while the Young Socialists called upon Mitterrand formally to condemn the Vichy regime.

Why Mitterrand allowed these revelations to be made is difficult to judge. 'His own political suicide' was the verdict of the Trotskyist Alain Krivine.[22] There was certainly an air of the confessions of a dying president at the end of his second Septennat. And yet, interviewed on television on 12 September 1994 Mitterrand, while acknowledging the revelations, obstinately refused to take full responsibility for his past actions. He dismissed his past as relating to 'petite histoire' rather than 'grande histoire'. He argued that his youthful right-wing politics had been determined by his Catholic, provincial, petty bourgeois background. He claimed that he had not known about Vichy's anti-Semitic legislation because he had been a prisoner of war at the time. He defended his relationship with Bousquet, pointing out that Bousquet had been acquitted and fully rehabilitated at the Liberation (developments in which he said he had had no part) and had been a leading light in business and Radical party circles after 1950. He was still reluctant to condemn Vichy, arguing that its crimes were the work of 'activist minorities', and that the Republic did not have to take responsibility for Vichy, since Vichy was not the Republic.

In all this, it may be argued, Mitterrand behaved no differently from other French citizens who had lived through the war years, and dealt with his personal history in the same way as the French dealt with their national history. For all the relevations, there was a reluctance by both the president and the French people to confront their past squarely and take responsibility for it. Only days before the publication of Péan's book France had indulged in a week of celebrating the 50th anniversary of the Liberation of Paris, as if the French nation had risen as one to expel the German Occupant. Not only was the reality altogether less heroic than the myth, but the myth had been woven specifically to conceal the reality. Increasingly, the myth appeared to be shabby and threadbare, and the finger of blame was pointed at certain individuals, some of them brought to justice. But to scapegoat a Bousquet or Papon, a Touvier or a Mitterrand, served only to allow the French to wriggle free once more of accepting responsibility themselves for the crimes and bad faith of the Vichy period.

4

Thirty Glorious, Twenty Inglorious Years

In 1979 Jean Fourastié, a former official of the economic planning agency, the Commissariat Général au Plan, published a book entitled *Les Trente glorieuses ou la Révolution invisible*.[1] The thirty glorious years, by analogy with the three glorious days of the July Revolution of 1830, were those of unparalleled growth and prosperity in France after the Second World War. He began by describing two villages, one backward and one developed, only to reveal that they were two snapshots of the same village, one in 1946, the other in 1975. He then set out a few eloquent statistics to highlight the changes in France as a whole between 1946 and 1975. The population grew from 40.5 to 52.6 million, and average life expectancy rose from 62 to 69 for men and from 67 to 77 for women. The proportion of those employed in agriculture fell from 36 to 10 per cent of the working population, while that employed in industry rose from 32 to 39 per cent and that employed in the service sector grew from 32 to 51 per cent. The standard of living measured in income per head of population increased (for a baseline of 100 in 1939) from 125 to 320, and the number of private cars in circulation rose from one million to over 15 million.

Baby boom and immigration

The 12 million by which the population of France increased in the thirty years between 1946 and 1975 equalled its growth during the previous century and a half. Part of it may be explained by a decline in mortality rates from 13.2 per 1,000 population in 1946 to 10.7 per 1,000 in 1964. More significant, however, was the sudden increase in the birth rate after 1943, from 14.6 per 1,000 population in 1938 to 21.4 per 1,000 in 1946, and still at 18.1 per 1,000 in 1964. Whereas on average women had two children in 1935, between 1942 and 1964 they had three. This baby boom was in no sense peculiar to France, but France had a long demographic history of a low birth rate which was now dramatically reversed. A number of explanations have been suggested. Government policy under Vichy and the Liberation, which provided generous family allowances in order to increase the population, contributed in some way, but the population was also reacting to changing circumstances. The trauma of Occupation and the joy of reunion at the Liberation highlighted the family as a source of comfort and security. The new generation of parents reacted against the small families of their own parents by having large ones themselves, taking advantage of a period of peace and prosperity. Those who were keen to improve their chances of social mobility, in the lower middle classes, tended to have fewest children, while those stuck at the bottom of society, such as agricultural and industrial labourers, and those who had arrived at the top, such as senior managers and the liberal professions, tended to have most.[2]

A third of the population growth, however, was explained not by an excess of births over deaths but by an excess of immigration over emigration. The number of foreigners in France increased from 1.7 million in 1946 to 3.4 million in 1975. As a proportion of the total population they rose from 4.3 to 6.5 per cent. To bring in foreigners was initially deliberate government policy, in order to help with the tasks of reconstruction by remedying the shortages of domestic labour. In 1946 a quarter of the foreign population was each provided by Italians and Poles, with Spaniards and Belgians following behind. By 1975, however, the largest foreign community was the Algerian, with 20 per cent, followed by the Portuguese, with 18 per cent. Foreigners by no means accounted for all the excess of immigration over emigration. Of the 3.5 million immigrants between 1950 and 1972, nearly two-fifths were repatriates from former French colonies, including a million *pieds-noirs* who returned from North Africa between 1962 and 1965, settling in the Paris region, the Midi, and Corsica. The issue of immigration, then, overlapped with that of the foreign population of France, but far from exhausted it.

Urbanization

The massive increase in population was reflected both in rapid urbanization and in a crisis of overcrowding, for the level of housing stock manifestly failed to keep pace with the demand from people moving to the cities. A quarter of the housing stock had been destroyed in the war, and post-war reconstruction initially gave priority to the rebuilding of ports, roads, and railways, rather than to new building. A survey of

1958 showed that 37 per cent of the population lived in overcrowded conditions, with three out of every ten families of four having only one or two rooms. Ninety per cent of homes in Paris had neither shower nor bath, and 73 per cent had no WC.[3] Immigrant populations tended to live in slum accommodation in the inner cities, such as the Goutte d'Or district of the eighteenth arrondissement of Paris, or furnished rooms in dingy hotels, more like bed-and-breakfasts. The poorest of the French populations, a sub-proletariat of the illiterate, unskilled and casually employed, now baptized the 'Fourth World', lived in shanty towns of breeze-blocks and corrugated iron such as that at Noisy, outside Paris, which sprang up in the 1950s.[4] Both poor French and immigrants also lived in 'transit camps' set up from the mid-1950s on the outskirts of Paris at Bagnolet, Créteil, Nanterre, and Stains, and also outside other towns like Lille, Mulhouse, and Toulon.

The French government launched more coherent building projects after the war, but the accommodation it built was either inadequate for needs or too expensive to rent, or both. In 1948 the ministry of reconstruction and town planning started to build cheap council houses, called after 1950 'habitations à loyer modéré' or HLM. However, competition for public resources meant that only 45,000 of these were built by 1953, and they were also designed for families with a reasonable income. In 1953 the ministry relaunched the programme, decreasing the size of the rooms and using prefabricated parts and reinforced concrete. It concentrated its efforts beyond the suburban villas developed between the wars, throwing up *grands ensembles* or high-rise estates of 8–10,000 units for a population of 30–40,000 people. The first of these estates was built at Sarcelles, to the north of Paris, in 1954, and 200 followed by 1964, of which 95 were in the Paris region. Some attempt was made to impose order on these developments after 1958, when they were concentrated on Zones d'Urbaniser à Priorité (Priority Urbanization Zones, or ZUPs). About 200 were established up and down the country, based on the notion of the separation of accommodation, industry, and offices, and increasing use of the car. Until the development of the hypermarket after 1963 they were severely lacking in services.

The planners of the Fifth Republic soon realised that the concentric growth of the large cities, especially Paris, was becoming suffocating, while the facilities in the outer suburbs were totally inadequate. They therefore invented a series of nine new towns, five of them to be strung out along the valley of the Seine, equipped with all the necessary facilities including industries and administrative offices and linked into

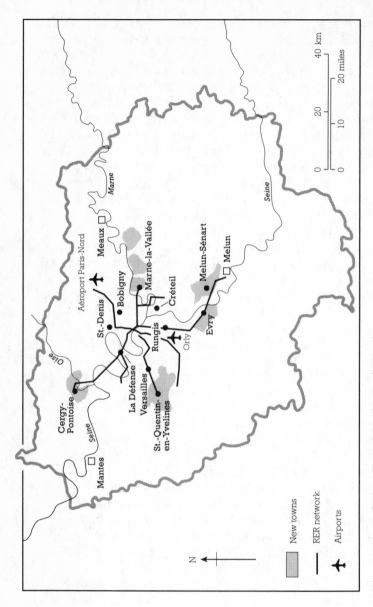

Fig. 4. Plan of New Towns in the Paris Region.

Sources: Pierre Barrère and Micheline Cassou Mounat, *Les Villes françaises* (Paris, Masson, 1980); Jacqueline Beaujeu-Garnier, *La France des villes*, i: *Le Bassin parisien* (Paris, La Documentation Française, 1978).

the transport system by a new high-speed rail link, the Réseau Express
Régional (Regional Express Network, or RER). The first, Cergy-Pontoise,
25 km. to the north of Paris, was begun in 1969 and had 140,000
inhabitants in 1987. The east–west artery of the RER was started in
1962 and the first section opened in 1969. The north–south line begun
in 1973 and the main intersection at Le Châtelet in central Paris opened
in 1977. Between 1978 and 1992 the number of passengers using the
RER every year doubled from 120 million to 237 million. And though the
new towns did not grow as fast as planned, the effect was to reduce
the population of central Paris by 800,000 by 1990, even if it remained the
densest metropolis in Europe with 200 inhabitants per hectare.

The Plan

The French economy at the Liberation was in a sorry state. Production
levels had plummeted, shortages were fuelling inflation, and capital
stock was outdated and in sore need of modernization. Meanwhile the
balance of power between employers, workers, and the state was
dramatically altered. Employers were weakened and discredited by
what amounted to the failure of capitalism in the 1930s and for having
profited from the Occupation by producing for the German war
economy. The working class, by contrast, emerged strongly organized
and vigorous after the war, legitimated by acts of Resistance, and
resorted to spontaneously taking over the industries of collaborating
capitalists, such as the mines of the Nord-Pas-de-Calais and the truck
factory of Berliet in Lyon. The state, for its part, was keen both to
assert itself against the power of the trusts and to limit workers' control,
and ensured this by a strategy of orderly nationalization.

Pierre Mendès France, who became minister of the national economy
at the Liberation, wanted to operate a threefold revolution. First, to
attack rampant inflation (and illegal wartime profits) by recalling all
bank notes and issuing new ones and fewer. Second, to nationalize key
sectors of the economy, notably banks and insurance companies, coal,
electricity, and petroleum, railways, shipping, air transport and lorries,
steel, and machine-tools. Third, he wanted to plan the economy
rationally, ironically by taking over the planning agency established by
Vichy in 1941 and plagiarizing its blueprints. Unfortunately he came up
against opposition from the Banque de France, the ministry of finance,
and the ministry of industrial production, and some wariness from de
Gaulle himself, particularly over his plan to fight inflation, and he
resigned in March 1945.

The attack on inflation went by the board. Some of the nationalization programme was carried out by the Constituent Assembly, notably that of the deposit (but not merchant) banks, coal, gas, and electricity, civil aviation, and the Renault car factory. The greatest success, however, was the planning of the French economy. Jean Monnet, a businessman who had been chief negotiator with the United States on economic and financial matters during the war, became head of the new Commissariat Général au Plan in January 1946. He ensured that his Commissariat—a compact body of about forty experts—was completely autonomous, responsible directly to the prime minister. He fought off take-over bids by the ministry of finance and national economy and published the first Plan by executive decree in January 1947, without consulting parliament. With government resources and self-financing by businesses so limited in the short term, he understood the necessity of American aid, which under the Marshall Plan provided a third of the investment planned between 1948 and 1950. Monnet was not a partisan of nationalization, not least because the Plan had to be sold to the Americans, but the existence of a nationalized sector in banking served to channel funds into heavy industry, the capacity of which to generate immediate profits was not guaranteed, while in turn the Plan ensured the success of the nationalized industries and banks. He disliked the *dirigisme* of Vichy and looked to set up a partnership between the state, business, and the labour unions, both to hammer out the details of the Plan in the twenty-four working parties set up under the Commissariat and to mobilize public opinion behind the Plan through the Conseil du Plan, which met under the chairmanship of the prime minister.

The first task of the Plan was reconstruction. The aim was to restore the 1938 level of production by the end of 1946 (in fact achieved in April 1947) and 1929 levels by mid-1948. Emphasis was placed on the basic sectors of coal, electricity, agricultural machinery, steel, and cement. The Plan benefited from total collaboration from the CGT and PCF, happy to accept American credits for their 'battle of production' in order to strengthen the independence of France. After the Communists were ejected from the government in May 1947 the CGT withdrew from the planning bodies, much to the regret of Monnet. Business leaders, who had initially cold-shouldered the Plan because they argued that it privileged the nationalized industries, now became the only collaborators of the government. The share of private investment rose, not least because from 1952 all American aid went into military projects under the auspices of NATO. The second Plan, of 1952–7, accordingly reflected the priorities of employers, putting the emphasis on the modernization of capital equipment and increasing the productivity of the labour force. More than a set of targets, it also helped to create a climate of expansion which, after the recession of 1952–3, was the opposite of the conventional wisdom of industrialists.

A burst of growth

The years down to 1973 saw a massive increase in investment in the French economy. It grew 4.9 per cent per annum in 1949–59, rising to 7.6 per cent in 1959–73. Investment, as a proportion of GNP, rose from 19 per cent in 1949 to 22 per cent in 1959 and to 27 per cent in 1973, a figure exceeded only by Japan and West Germany.[5]

The economy was also stimulated by the opening up of trade. France had traditionally been a protectionist country, and a minor player in the world economy. She kept a cosy two-way trade with her empire largely to herself. One advantage of this relative isolation was that the trade depression of the 1930s hit France less seriously than many other European countries. But 1 January 1959 saw the inauguration of the European Economic Community, and the competitiveness of France was put decisively to the test. In September 1958 de Gaulle put Jacques Rueff, a theorist of economic liberalism and former adviser to Raymond Poincaré, in charge of a committee to prepare France's entry into the Common Market. He believed firmly that the future growth of the French economy depended on throwing open her frontiers and increasing international trade. By building up a healthy balance of trade in 1959–62 France was encouraged to throw away her protective crutches much faster than the timetable required. She sharpened her competitive position by devaluing the franc by 17.5 per cent at the same time as replacing 100 old francs by one new franc, close in value to the Swiss franc or German mark. And lest there be a rush of imports once the barriers came down, the Rueff plan included tax increases and cuts in government expenditure.

The gamble paid off. The volume of France's foreign trade, which had grown by 4 per cent per annum before the war and by 6.5 per cent between 1949 and 1959, increased by 10.8 per cent in the period 1959–74. France became the fourth largest exporting country by 1964. Her exports, standing at 10 per cent of her GDP in 1958, rose to 17 per cent in 1970. In all this, Europe played a key part. While the proportion of French exports to the franc zone (essentially her former empire) fell from 30 per cent to 10 per cent in the period 1958–70, her exports to Europe increased from 10 per cent to 50 per cent of the total.

TABLE I. *Comparative Annual Rates of Growth, 1949–1971*

	1949–59	1959/60–1970/1
France	4.5	5.8
West Germany	7.4	4.9
Italy	5.9	5.5
UK	2.4	2.9
USA	3.3	3.9
Canada	4.2	4.9
Japan		11.1

Source: Fernand Braudel and Ernest Labrousse (eds.), *Histoire économique et sociale de la France*, iv/3 (Paris, PUF, 1982), 1012.

The impact of high investment and expanding trade on France's rate of growth was very impressive. Between 1949 and 1959 her economy grew at the rate of 4.5 per cent per annum. This was faster than that of the USA (3.3) or the UK (2.4), but substantially behind that of West Germany, which grew during her economic miracle at 7.4 per cent per annum. After her entry into the Common Market, France's economic performance stepped up a gear. Between 1959–60 and 1970–1, the British economy (then outside the EEC) expanded by only 2.9 per cent per annum, and the American by 3.9 per cent, while the West German economy fell back to an increase of 4.9 per cent. The French economy, by contrast, surged ahead to an expansion of 5.8 per cent per annum, second only to that of Japan, 11.1 per cent. Unemployment in France was reduced to a vestigial 1–2 per cent of the working population in the period 1950–70. The only major problem was the rate of inflation, which was the highest in Europe in the early 1960s, although in the light of later developments 4 per cent inflation between 1959 and 1969 did not seem outrageous.

The end of the Peasantry?

The economic transformation of France cannot be expressed in statistics of growth alone. The economic and social structure of France changed beyond all recognition within the space of a generation. This can be plotted in a number of ways: as the 'tertiarization' of the economy from agriculture and industry to an expanding service sector; as the displacement of farmers, artisans, shopkeepers, and businessmen who owned their own businesses by a new class of salaried managers who did not; or as the disappearance of the peasantry, the rise of a new working class, and the hegemony of the *cadres*.

France had long been a country dominated by its peasantry and its rural population. At the beginning of the Second World War 45 per cent of the population lived in rural communes, defined as having fewer than 2,000 inhabitants. However, agriculture changed more in the thirty years after the war than it had changed in the previous century and a half. It modernized and industrialized at a fantastic rate, and agricultural production doubled between 1945 and 1974. This enabled France to feed her rapidly growing population and even to export significant amounts of foodstuffs; the irony of it was that success in production drove down prices and spelled catastrophe for a large proportion of the farming population which was left high, dry, and surplus to requirements.

The great discovery of the immediate post-war years was the tractor. In 1946 there were 20,000 tractors in France; in 1965 over a million, equipping over half of French farms. The shortages and high farm prices down to 1948 meant that farmers could pay for their tractors out of large profits. Subsequently, the cost not only of equipment but also of fertilizer and animal feed caused farmers to beat a path to the Crédit Agricole, of which over two thirds were members by 1959. The application of science to agriculture threatened to make farmers slaves to the tractor companies, laboratory experts, and peddlers of new chemicals and crop strains, and drove their sons to acquire a more sophisticated agricultural training, either in agricultural colleges or in rural apprenticeship schemes. This provoked something of a conflict of generations in the countryside. Educated and dynamic young militants joined a branch of Catholic Action, the Jeunesse Agricole Chrétienne (Christian Farming Youth Movement, or JAC), 440,000 strong in 1954. This took over the Cercle National des Jeunes Agriculteurs (National Circle of Young Farmers, or CNJA), making it into a highly organized arm of the young farmers, and in 1957 established a foothold in the dominant union, the Fédération Nationale des Syndicats d'Exploitants Agricoles (National Federation of Farmers' Unions, or FNSEA).

The costs of farming obliged farmers to abandon mixed farming and autarky for market orientation and specialization. This opened up another conflict, between those who had successfully adapted, those who were seeking to adapt, and those who had neither the inclination nor the ability to adapt. There was some overlap here with the generational and institutional conflicts already described. At one end of the scale were the large capitalist wheat-farmers of the Paris basin and the north of France, who were highly commercialized, industrialized, able to practise economies of scale, and who controlled the FNSEA and access to power. At the other end were traditional peasant farmers, who practised a mixture of arable and pasture on the poor soils of the Alps, Pyrenees, and Massif Central, for local markets and behind protective walls thrown up under the Third Republic. Some of them founded a Mouvement de Défense des Exploitants Familiaux, close to the CGT, in April 1959, and tended to vote Communist. In the middle were young farmers of Brittany and Normandy, the Rhône valley, and Languedoc who ran the CNJA. They were short of land, monopolized as it was by capitalist farmers and aged peasants. They sought a way out in high value-added cash crops such as dairy products, pork, poultry, wine, fruit, and vegetables.

Conflict was generated by the falling agricultural prices that resulted from overproduction and was articulated around the agricultural policies of the Fourth and Fifth Republics. The FNSEA, representing the large farmers of northern France, secured the election of about 100 deputies favourable to its interests in 1951, and effectively controlled the ministry of agriculture between 1951 and 1956. It demanded price supports, backed up its demands with road-blocks in October 1953, and forced the government to concede marketing boards to underpin milk, potato, and wine prices. Though it had less direct influence in government after 1956, the FNSEA sponsored widespread demonstrations in May 1956, and in September 1957 the Radical-led government was obliged to accept the indexation of farm prices to price rises in general.

On 28 December 1958 the Fifth Republic, in the context of the Rueff plan for a highly competitive economy, abolished the indexation of agricultural prices. The FNSEA, embarrassed for having supported de Gaulle, organized a mass meeting of 30,000 at Amiens in February 1960, which erupted into violence, and pressed for an emergency recall of parliament, which the government refused. The government's policy for agriculture was not price support but restructuring, a shift to larger, more efficient farms producing at low cost, which would mean driving old, small, and inefficient farmers off the land. This strategy, as it happened, fitted in with that of the young farmers of the JAC and CNJA, some of whom had been elected to parliament as MRP deputies in 1958 and 4,000 of whom were elected rural mayors in 1959. A law on the orientation of agriculture in August 1960 established this new direction, but nothing was then done to enforce it. Infuriated, the young farmers rioted in May 1961. The movement started among the market gardeners and poultry-farmers of Brittany, where truckloads of potatoes were dumped in town halls, ballot boxes were burned, and the subprefecture of Morlaix occupied. It spread to the wine-growers, fruit-growers, and market gardeners of Languedoc and to Normandy and Aquitaine. Roads and railways were blocked, telephone lines were brought down, and prime minister Debré was burned in effigy. It dragged along some of the traditional peasants but did not move the big cereal-growers of northern France. Eventually a complementary law of August 1962 gave the young farmers what they wanted: an indemnity scheme to facilitate the retirement of old farmers, and semi-public bodies with a right to pre-empt farms that thus came onto the market, keeping them out of the hands of the speculators and making them available to enterprising young farmers.

The effect of these reforms and the harsher economic climate was dramatic. Between 1954 and 1974 the number of farms fell from 2.3 to 1.3 million. The working agricultural population (including agricultural workers and retired farmers), as a proportion of the total working population, fell from 27 per cent in 1954 to 20 per cent in 1962 and under 10 per cent in 1975. The number of farmers declined from 4 million

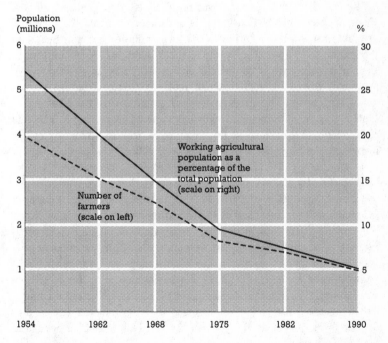

FIG. 5. Decline of Number of Farmers and Proportion of Working Agricultural Population, 1954–1990.

Source: *L'Etat de la France 94–95* (Paris, Éditions La Découverte, 1994).

in 1954 to 1.6 million in 1975. Of 32,655 male farmers who left farming businesses to find work in 1972–3, 40 per cent became industrial workers, 30 per cent agricultural workers, and 19 per cent workers in the tertiary sector; the percentages for 13,008 female farmers who moved were quite different, respectively 17, 19, and 54.[6] At the same time the extent of the changes should be put in perspective. Price support, no longer provided by the French government, was provided by the Common Agricultural Policy set up in 1961–6 as a condition of France's entry into the EEC, and France received 40 per cent of CAP funding for this in 1971. Moreover, the average size of French farms increased from 15 hectares in 1954 to 28 hectares in 1988—a significant but not revolutionary change—so that France remained very much a country of small and medium farmers.

The new working class

If the traditional French peasantry was one disappearing species, the traditional French proletariat was another. By this is meant not the working class in general but workers in those sectors that characterized the first Industrial Revolution: textiles, coalmining, iron and steel, and the railways. In 1947, for example, there were 330,000 coalminers in France; in 1980 only 33,000 remained. Sociologically, this generation of industrial workers, which reached maturity between the 1930s and 1950s, has become known as the 'unique generation' of the proletariat.[7] Unlike the previous generation, which was often recruited from the countryside, the former was born into the working class and grew up in defined working-class communities such as the textile conurbation of Lille–Roubaix–Tourcoing, the coalfields of the Nord-Pas-de-Calais, the steel towns of Lorraine and Villeneuve-Saint-Georges, a railwaymen's town on the outskirts of Paris. Part of the community was composed of immigrant workers, such as Polish miners and Italian steel workers, but they had become effectively amalgamated with their French comrades. Historically, they had been shaped by the same powerful experiences of the Popular Front, the German Occupation, and the strikes of 1947. With a strongly developed class-consciousness, they were organized in the Communist-dominated CGT and the Communist party, which at the Liberation took power in working-class towns and displaced employers as the provider of housing, social services, and cultural facilities.

This traditional working class was threatened after the war by new technology permitting the substitution of capital for labour, by new sources of energy, such as hydro-electricity, gas, and nuclear power, and by subjection to international competition, particularly within the EEC. In industries that were not concentrated into large units of production, such as textiles, the work-force gave up the ghost without much of a fight. Railwaymen, whose numbers declined from 46,000 in 1949 to 35,000 in 1961 as a result of electrification and automation, struck in June 1962, but to no long-term effect. The miners, victims of the plan of the minister of industry and commerce Jean-Marcel Jean-neney to close down pits and streamline wages and conditions, sustained a five-week strike in March–April 1963, and were supported by steel workers, gas, electricity, rail, and metro workers. The government was forced to concede an 11 per cent pay rise and fourth week of paid holiday, but the rundown of the pits continued unabated. Unlike after 1973, however, those who had been laid off in traditional industries were rarely left without work. Most simply transferred to modern and booming sectors of the economy, such as the seaport steelworks of Dunkerque and Fos-sur-Mer, near Marseille, and the engineering, chemical, electrical, and automobile industries that were shaping a new working class.

The new generation of the working class that appeared in the 1960s was the result of a strategy to decentralize much of French industry, in pursuit of cheap sites and cheap labour and a more equitable balance of resources and employment between Paris and the provinces. It was once again heavily recruited from the countryside among farmers or their sons seeking alternative work, or from local artisans and tradesmen made bankrupt by the collapse of the rural economy. It tended to live on housing estates in ZUPs rather than in inner-city districts, and was less unionized than its predecessor. It included a greater proportion of women, who had been heavily represented in traditional sectors like textiles but much less so in the men's world of coal and steel. It included a high proportion of immigrants, notably the new immigrants from North Africa. Three-quarters of immigrants worked in unskilled or manual occupations, while 12,000 workers or 39 per cent of total work-force at the Billancourt Renault works in 1969 were immigrants. The new working class was less skilled, as the scientific organization of work, sometimes known as Taylorism, especially the introduction of automation, the assembly line, and shift system, displaced workers with a trade (*ouvriers professionnels* or OP) in favour of semi-skilled workers (*ouvriers spécialisés* or OS). In the Renault car factories between 1965 and 1969 alone, the proportion of OP declined from 27 to 20 per cent of the personnel, and that of unskilled labourers and porters from 3 to 2 per cent, while that of OS rose from 69 to 78 per cent.[8] The intellectual tasks done by skilled workers were increasingly done by technicians, whose numbers multiplied in France from 193,000 in 1954 to 531,000 in 1968.

It was hoped by employers that the new working class would be more docile and amenable than the old. When the Bull computer company established a factory in Vendôme in 1953, its employees were overwhelmingly women, 10 per cent of them aged 14–16, and were bussed daily out of the countryside. In Caen, the traditional working class was represented by the steel workers of the Société Métallurgique de Normandie (SMN), masculine and solidly organized by the CGT. Between 1959 and 1965, however, the town saw the arrival of Renault Industrial Vehicles (RVI), Citroën, Radiotechnique, belonging to Philips, Blaupunkt, which made car radios, Jaeger, and Moulinex. This created 15,000 jobs on top of the 5,000 at SMN, and a new, mainly female, rural, and young working class. Far from being docile, however, the new work-force was infiltrated by the CFDT union, far more radical than the CGT, and a strike at the RVI factory in January 1968 rapidly spread to Radiotechnique, Jaeger, and Moulinex. Young workers clashed with police in the town during the night of 28 January. It has been asserted that the revolution of 1968 started at that moment in Caen. In any case, it was not the traditional working class that struck in May 1968 but the new working class of the engineering, electronics, aircraft, automobile, and chemical factories.

The new middle classes: the *cadres*

The development of large-scale industry, the industrialization of commerce as supermarkets drove out small shopkeepers, and the growth of bureaucracy both public and private spawned a new phenomenon in French society: the *cadre*. The decline of the bourgeois who owned his own business, large or small, and handed on the family fortune or *patrimoine* to the next generation was mirrored by the rise of a salaried middle class, in theory meritocratic (although 71 per cent of *cadres* in 1977 were children of *cadres*) and dependent on the income from a highly paid job rather than on the inheritance of any family fortune. Statistically, self-employed business people, shopkeepers, and artisans declined as a proportion of the working population between 1954 and 1975 from 13 to 8 per cent, while the proportion of higher-level *cadres* rose from 3 to 8 per cent and that of middle-level *cadres* rose from 9 to 17 per cent.[9]

The *cadre* slotted into the business world as a manager, between owner and worker. He or she was defined by two qualities: first, by education and intellectual capability and second, by a degree of responsibility or delegated authority within the business to supervise and organize work. Ideologically, the *cadre* sought a middle way between capitalist anarchy and collectivist tyranny, and had been given a place in the Charter of Labour of the Vichy government between organized capital and organized labour. Arbitrators of social peace, they were also heralds of modernization and believed in the scientific organization of economic life, developed first to increase productivity, then to improve marketing, and finally to perfect financial controls.

The model for the *cadre* was the *ingénieur*, that product of one of France's élite technical schools, the *grandes écoles*, skilled in mathematics and considered to be the élite of economic life. Not that all *cadres* were as well educated as *ingénieurs*: 70 per cent of male and 73 per cent of female *ingénieurs* had degrees in 1981, as against 32 per cent male and 33 per cent female senior managers. But those expert in maths and technology were not always good managers of people, and in the 1950s and 1960s the French discovered the American science of management, not least through missions to the USA under the auspices of the Marshall Plan. Management acquired its own autonomy and status, and *L'Express*, founded in 1953, became known as the *cadres'* magazine. In time the *cadres* came to marginalize the *ingénieurs*, not only in numbers but also in clout; in 1954 the INSEE population surveys placed *cadres supérieurs* in the same socio-economic category as *ingénieurs*, teachers in secondary and higher education, and the liberal professions, while middle managers were grouped with technicians, primary school teachers, and medical and social service professionals other than doctors. In 1982 INSEE refined the terminology, placing senior managers with the 'higher intellectual professions' and middle managers with the so-called 'intermediary professions', but the hierarchy remained the same.

A last distinction that was not always so clear was between the public and private sector. Between 1962 and 1982 INSEE obliterated the distinction between *cadres* in the private and those in the public sector. There were two main reasons for this. The *cadres* modelled themselves on the *ingénieur* as far as economic status was concerned, but on the civil service in so far as they wanted a hierarchical profession, with promotion according to seniority as well as capability, and a scale of steadily rising salaries to compensate for lack of profits or family fortune. In addition, civil service and business collaborated closely in the drafting of economic plans; the scope of nationalized industries was wide after the Liberation and private firms also became large and bureaucratic; there was even a well-trodden path of moving from an early career in the administration, close to the centres of power, to a later, more lucrative career in the business world. All this added up to embedding the *cadres* firmly in the public and private governing élite in France.

The oil crisis and other problems

On 20 December 1973 Jean Fourastié published an article in *Le Figaro* entitled 'The End of the Easy Times'. This proved remarkably prophetic. The thirty glorious years came to a sudden end with the oil crisis of December 1973, when the OPEC countries quadrupled the price of oil exports during the Arab–Israeli war. Since France imported three-quarters of her energy, the impact was particularly brutal. Inflation went up to 14 per cent in 1974, the balance of trade went sharply into deficit, and increased costs pushed industry into recession.

The oil crisis came on top of another dramatic event. On 15 August 1971, in an attempt to kick-start the American economy, Richard Nixon suspended the convertibility of the dollar into gold. Since other major currencies were pegged to the value of the dollar, this had the effect of ending fixed exchange rates and throwing international money and trading markets into confusion. It ended the long period of monetary stability established at Bretton Woods in 1944. Some order was restored to money markets with the establishment of the 'snake' in April 1972, linking European currencies, but the effect of the oil crisis drove France out of the snake between January 1974 and May 1975, and again in March 1976. Not until the negotiation of the European Monetary System (EMS) by France and Germany in 1979 was genuine stability restored to money markets and France able to pursue her policy of the 'strong franc'.

Closer to home, the era of planning inaugurated by Jean Monnet in 1946 effectively came to an end after the Fourth Plan of 1959–65. Plans continued to be drawn up, but they had less and less bite. The traditional departments of state reasserted their dominance, and businesses reasserted their autonomy. The relative weakness of the business world in the immediate post-war period, which had allowed scope for state planning, no longer obtained once business had recovered and restructured itself. Renault, for example, insisted on launching the 4CV, though the planners wanted it to specialize in heavy goods vehicles. The

1960s saw the re-emergence of finance capitalism and the formation of large industrial groups and international conglomerates which were much less amenable to bullying by the state. Two important holding companies were founded, the Compagnie de Suez, heavily compensated by Egypt for the loss of its interest in the Canal, in 1966, and the Compagnie Financière de Paris et des Pays-Bas (Paribas) in 1968. These acquired interests in banks, insurance companies, and industrial firms, and encouraged mergers in order to reduce domestic competition, sustain international competition, and safeguard their investments. Thus Suez prompted the fusion of the chemical companies Saint-Gobain and Pont-à-Mousson in 1970. The chemical industry became dominated by three groups, Saint-Gobain-Pont-à-Mousson, Rhône-Poulenc, and Pechiney-Ugine-Kuhlmann, the automobile industry by Renault, Citroën, Peugeot, and Simca, the steel industry by Wendel-Sidélor, Denain-Nord-Est-Longwy and Creusot-Loire, while a computer group CNII was founded in 1967. On the world scene, however, these groups found it difficult to compete. Renault, the largest firm in France, was only the twenty-second largest firm in the world, the steel groups were very modest indeed, and CNII was a midget in comparison with IBM.

More serious than all of this, however, were certain structural problems in French industry that made it slow to modernize and increasingly uncompetitive in the world economy. The first problem was the high relative cost of labour and social security. The Grenelle accords of 1968 had been a great victory for the working class but illustrated the tendency of the French government to give in to labour demands for the sake of social peace. The minimum wage or SMIG (Salaire Minimum Interprofessionnel Garanti), established in 1950, which from 1952 was indexed to prices and had fallen behind the average wage between 1955 and 1967, was subsequently linked to the average wage as well as to prices, renamed the SMIC (Salaire Minimum Interprofessionnel de Croissance) in 1970, and rose more rapidly than the average wage between 1968 and 1975. Meanwhile French expenditure on social security, which as a proportion of GNP was less than that of Germany and Belgium in 1970, exceeded that of Germany and Belgium in 1980, and equalled that of Denmark. Only the Netherlands and Sweden spent significantly more, Italy, Great Britain, and the USA always substantially less.[10]

The growing proportion of business and national incomes spent on wages and social security charges meant less spent on investment. Investment in France stagnated between 1974 and 1980 and as a proportion of GNP fell from 27 per cent in 1973 to 21 per cent in 1980. The productivity of capital, which had been on the decline since 1963, fell dramatically after 1973. The rate of economic growth, which had been 5.8 between 1960 and 1973 (the same figure as between 1959–60 and 1970–1), actually went into reverse (−0.3) in 1975. Between 1973 and 1979 it was 2.8 per cent, which was above the average for the European Community and the United States but lower than that of Japan. Between 1979 and 1984 the growth rate in France plummeted to 1.1 per cent per annum. It improved to 2.2 per cent between 1985 and 1987 and to 4.4 per cent in 1988 and 1989. Overall, between 1979 and 1990, it was 2.2 per cent, fractionally higher than that in Great Britain and Germany, but lower than that in Italy, the United States, and Japan. After 1990, however, growth slumped again. The slowing down was not as severe as in the United States in 1991, let alone as in Great Britain in 1991 and 1992, but France went into recession in 1993, along with Germany and Italy, just as the United States and Great Britain and the USA were recovering.

Another sign of the unhealthiness of the French economy was the balance of payments. Since France imported so much of its energy the oil crisis was devastating, but the trade surplus had been decreasing in the 1960s and disappeared in 1970. More and more, France had the profile of a developing country. Its imports of manufactured goods increased by 9.6 per cent per annum between 1967 and 1987. The money-spinners in the 1980s were arms sales, the export of farm produce, and tourism. Between them, tourism and farm produce earned enough to pay the oil bill, while as much was made by farm-produce exports as by arms sales. Inflation continued to rise. Between 1973 and 1978 the inflation rate in France was 11.0 per cent, lower than the 15.8

TABLE 2. *Comparative Annual Rates of Growth, 1960–1993*

	1960–73	1973–9	1979–90	1991	1992	1993
France	5.8	2.8	2.2	0.7	1.4	−0.7
Germany	4.4	2.3	2.0	4.6	1.2	−0.9
Italy	5.3	2.6	2.5	1.3	0.5	−0.5
UK	3.2	1.5	2.1	−2.3	−0.4	1.9
EEC	4.8	2.4	2.3	1.5	1.1	−0.5
USA	3.9	2.6	3.0	−0.7	2.6	2.9
Japan	10.6	3.6	4.2	4.0	1.3	0
OECD	4.7	2.6	2.6	0.8	1.7	1.1

Source: L'État de la France, 94–95, 391.

TABLE 3. *Comparative Rates of Unemployment, 1970–1993*

	1970	1975	1980	1985	1990	1993
France	2.5	4.0	6.3	10.2	9.0	11.9
West Germany	0.6	4.0	3.2	8.0	6.2	8.9
Italy	5.3	5.8	7.5	9.6	10.3	10.2
UK	2.2	3.2	5.6	11.5	5.5	10.3
EEC	2.5	4.2	6.2	8.0	8.5	10.7
USA	4.8	8.3	7.0	7.1	5.4	6.9
Canada	5.6	6.9	7.4	10.4	8.1	11.2
Japan	1.1	1.9	2.0	2.6	2.1	2.5
OECD	3.4	5.4	5.9	11.0	6.1	8.2

Source: *L'État de la France, 94–95*, 403.

per cent in Great Britain, but much higher than the 5.0 per cent in West Germany. The second oil crisis, provoked by the collapse of oil production in Iran and Iraq, drove inflation up to 13 and 14 per cent in France in 1979–81. Meanwhile unemployment, which had been negligible in the 1960s, reached the 1 million mark in 1975, 1.5 million in 1980, 2.0 million in 1982, 2.6 million in 1986, 2.8 million in 1991, and 3.4 million at the end of 1993. The unemployment rate in France was always above that in the EEC as a whole after 1980, and though the unemployment rate in Great Britain was higher than that in France in 1985, subsequently it overtook that in Great Britain and remained above it. Finally, as we shall see, inflation was brought under control, but the other great challenge, to increase industrial competitiveness and economic growth and reduce unemployment, was not met.

Government policy: deflation or reflation?

French governments in this period had two policy objectives: to control inflation and to increase industrial competitiveness and economic growth and reduce unemployment. In some sense, these were complementary targets. To control inflation would make exports cheaper and therefore more competitive. But it did not always work like that, and where the targets clashed, all too often the war against inflation took precedence over the battle to modernize.

Deflation had been a priority for Giscard d'Estaing when he was finance minister in 1963, and it was once again a priority when, as president, he launched a stabilization plan with his prime minister, Raymond Barre, who he proclaimed to be the best economist in the country. The idea was to bring down prices, which would allow wages and social costs to fall, and reduce public expenditure. In March 1979 France entered the European Monetary System, which was designed to produce a strong franc and keep down inflation. Businesses rather than the government were entrusted with the task of restoring prosperity and growth, with their wage bill lower, profits restored, and corporate taxes reduced. After 1978 liberalism was the order of the day. Price controls on businesses were relaxed and their profits increased by 9 per cent in real terms (while the purchasing power of workers increased by only 2 per cent), creating a surplus for investment. The strategy, however, failed to work. Increased profits were not invested, and Barre's forecasts for growth proved wildly optimistic. The strong franc overpriced French goods, which lost a substantial share of the German market between 1979 and 1982. The goal of price control and the means of liberalizing prices were at odds with each other, and in any case the anti-inflationary forces were completely routed by the second oil crisis. Lastly, unemployment rose from 1 million or 4.0 per cent of the working population in 1975 to 1.5 million or 6.3 per cent in 1980.

In the winter of 1980–1, preceding the presidential election, Barre reverted to a policy of reflation. It was, however, the incoming Socialist president, François Mitterrand, and his government who dramatically reversed the strategy of deflation and applied Keynesian methods to increase domestic demand, boost exports, and stimulate growth. The minimum wage was increased by 10 per cent, child benefit by 50 per cent, and old-age pensions by 62 per cent. About 150,000 new jobs were created in the public sector, funds provided to build 50,000 new homes, and plans made to employ 650,000 young people. The working week was reduced and the retirement age lowered to 60 in order to spread employment. The government was prepared to allow a budget deficit, but new taxes were imposed on the rich, notably the Impôt sur les Grandes Fortunes (IGF). A programme of nationalizations was executed, involving the holding companies Paribas and Suez, 36 private banks, and 9 industrial groups, including the chemical groups Pechiney-Ugine-Kuhlmann, Saint-Gobain, and Rhône-Poulenc, the electrical groups Thomson-Brandt and the Compagnie Générale d'Éléctricité (CGE), and the steel conglomerates Usinor and Sacilor; a 51 per cent stake was also taken in the armaments companies Dassault and Matra. Nationalization was not only an article of Socialist faith, it was designed to increase public investment in industry and to restructure and modernize it. Thus the electronics industry was divided between CGE, to concentrate on telecommunications, and Thomson, to concentrate on televisions and the like, while Pechiney was relieved of steel and chemicals in order to concentrate on aluminium. The rationalization was continued with a ninth Plan, drafted for 1984–8, less to dictate to business than to reduce uncertainty. The competitiveness of industry was enhanced, finally, by three successive devaluations of the franc, in October 1981, June 1982, and March 1983.

The strategy of reflation, however, was just as bad as that of deflation, if not worse. It was predicated on forecasts that the world economy would pick up and sustain French investment and production, but the forecasts were mistaken and there was a downturn in world trade in 1982. There was a balance of payments disaster, because France's industrial base was so weak that the new demand in the French economy, unable to satisfy itself domestically, sucked in foreign imports. Inflation, meanwhile, was rampant, reaching 14 per cent in 1981 and 1982, while in West Germany it fell from 6.0 to 3.6 per cent. Growth, predicted to be 3.3 per cent in 1982, turned out to be a mere 2 per cent. Unemployment, far from being reduced, actually rose from 1.5 million and 6.3 per cent of the working population in 1980 to 2.0 million and 8.6 per cent in 1982.

In response to this the Socialist government undertook the biggest U-turn in recent French economic history. The great difference economically was not between Giscard's liberalism and Mitterrand's socialism but between socialism before March 1983 and socialism afterwards. The switch to a deflationary policy began with a six-month price and wage freeze imposed in June 1982. In March 1983 the franc was devalued for the third time, but a more radical devaluation, which would have forced France to leave the EMS, was ruled out. Committed to the EMS and the war against inflation, the government introduced a host of tax increases and public expenditure cuts. Pierre Bérégovoy, who had previously favoured leaving the EMS, changed his mind and was moved to the ministry of social solidarity to slash the social security budget. Laurent Fabius, the industry minister, took the first steps of privatization by allowing private capital up to a 25 per cent stake in the nationalised industries, and allowing the privatization of subsidiaries of firms like Pechiney. In addition, he accepted the inevitability of significant lay-offs in the coal, steel, and automobile industries. The main beneficiaries of the change were businesses, as the Socialists now reverted to the liberal strategy of 1978, reducing their corporate taxes in the hope that they would invest profits in training and jobs. This time the policy began to bite. Inflation was brought down to 5 per cent in 1985. The balance of payments deficit was reduced. On the other hand, nothing was done for growth, which was 1.9 per cent in 1985, while at the same time unemployment rose to 10.2 per cent of the working population.

This liberal policy was continued under the conservative government of Jacques Chirac and his finance minister Édouard Balladur in 1986–8. Prices, some of which had been regulated since 1945, were completely deregulated in December 1986, and Paribas, Suez, the Société Générale, and twelve major groups including Saint-Gobain, GCE, and Matra, nationalized by the Socialists, were privatized. Prices were not driven up, because of the marked decline of world oil prices after 1985; in fact the rate of inflation in France was 2.2 per cent in 1986, and 3 per cent in 1987 and 1988. Growth improved to 2.2 per cent in 1986 and 1987, and the level of unemployment stabilized. On the other hand, the balance of payments record was still bad in 1986 and deteriorated in 1987. The world economy improved in the second half of the 1980s, but the fragile industrial base of France meant that it was not exports that responded, but imports.

The dominant economic figure in the second Socialist era of 1988–93 was Pierre Bérégovoy. Finance minister from 1984 to 1986 and again from 1988 to 1992, then prime minister in 1992–3, he made virtues of the strong franc and 'competitive disinflation'. Nicknamed 'the father of austerity', a 'left-wing Pinay', or 'Monsieur Périgovoy' by analogy with M. Périgot, president of the employers' union, he was more deflationary and liberal than the liberals. He reduced public expenditure, in defiance of Prime Minister Rocard's inclination to spend, and reduced corporation tax on businesses. He brought the inflation rate down to an average of 2.7 per cent in 1988–93, lower even than that of Germany after 1991. The balance of payments improved, largely as a result of the collapse of internal demand and the demand from newly united Germany, and actually moved into credit in 1992. Growth climbed to 4.4 per cent in 1988 and 1989, before declining to 2.2 in 1990, 0.7 per cent in 1991, 1.4 per cent in 1992 and going into recession in 1993. The rate of unemployment, meanwhile, fell to 9.0 per cent of the working population in 1990 but rose to 10.2 per cent in 1992.

Édouard Balladur, who became prime minister in March 1993, gave the deflationary and liberal strategy an additional twist. Whereas Great Britain had been driven out of the EMS by a tide of speculation in November 1992, Balladur maintained the strong franc within the EMS, with the help of the Bundesbank, in defiance of the speculation launched against it in July 1993. As France sank into recession, his priority was not to boost the economy but to restore sound public finances by requiring 'sacrifices' from the mass of the population. Public-sector wages and family allowances were frozen, medical reimbursements were reduced, and the SMIC decoupled from the average wage. The tax on petrol, meanwhile, was doubled. Economic recovery, he argued, would come from the private sector. With this in mind, corporate taxes and income tax were reduced, and greater 'flexibility' allowed to employers in the hiring and firing of labour. The economy, which went into recession in 1993 (– 0.7 per cent growth), recovered to 2 per cent in 1994. The balance of payments remained in credit for a second year in 1993 and inflation was virtually wiped out. All this, however, was reflected in a lack of demand in the economy, caused largely by massive unemployment, which rose to 11.8 per cent of the working population at the end of 1993 and to 12.6 per cent in 1994.

Modernization and de-industrialization

Despite the emphasis on deflation and liberalization, it would be wrong to assume that French governments had no policy of modernization for industry. The problem here was that some sectors of the economy were more susceptible to modernization than others, and that increased competitiveness required difficult decisions about laying off large numbers of workers. While this made sense for individual firms or industries, it threatened to reduce demand in the economy as a whole and impose heavy demands on the state budget.

The strategy launched by Giscard d'Estaing was the so-called *politique de créneaux*, or target strategy, to privilege leading sectors of the French economy to compete effectively in world markets and to prevent the penetration of those sectors of the French market by American or Japanese products. Those leading sectors included the arms industry, characterized by Dassault's Mirage fighters and Aérospatiale's Exocet missile, which was worth 62 billion francs of exports in 1984. The nuclear industry was developed with especial commitment after the oil crisis of 1973. By 1985 nuclear power generated 65 per cent of French electricity (compared to 31 per cent of electricity in West Germany and 19 per cent in the UK) and exported current to Italy and Switzerland. The aeronautical industry was able to rival Boeing with the Airbus, developed by Aérospatiale in conjunction with the German, Italian, and British aircraft industries. The space programme concentrated on satellites launched commercially by Ariane after 1979. The transport industry thrived on contracts to built metros and high-speed railways at home and abroad. Other leading sectors were not so successful. The telecommunications industry developed the Minitel electronic directory, but France lagged behind with the laying of cable and erection of satellite dishes and the market was invaded by the Americans with data transmission and Asians with faxes and cordless phones. In the computer industry, CII was taken over by Honeywell Bull in 1976, but Bull was still a fraction of the size of IBM, with which it had to ally in order to keep up with technological change. In the mass electronics industry Thomson was the fourth largest firm in the world, but while it was strong in colour television it failed to figure in the walkman, video-recorder, video camera, and video game markets, so that 90 per cent of the domestic market in 1991 was imported from South Korea, Taiwan, Hong Kong, and Singapore.

The disadvantages of this strategy was that it privileged those sectors where the state played an important part taking initiatives and winning contracts against those where it did not, large companies with access to international markets against those which did not, and high-tech and capital-intensive industries against more traditional ones. The Socialists took the view after 1981 that there were no outdated sectors, only outdated technologies. They balanced the *politique de créneaux* with a *politique de filières*, or diffusion strategy, looking to revolutionize all sectors of industry from the top downwards, notably by the spread of information technology. They thus concentrated three-quarters of investment in the coalmining, steel, and shipbuilding industries. This was entirely praiseworthy, and intended to save the working class that had elected them. But the traditional sectors simply could not compete with Japan and the emerging markets of South-East Asia. The steel basins of Longwy and Valenciennes had been on the verge of insurrection against planned closures from December 1978 to May 1979. President Mitterrand visited Longwy in October 1981 and called it the symbol of the failure of a policy. But the closures continued after the socialist U-turn, and 60,000 steel workers demonstrated in Paris on 13 April 1984, backed by the local press proclaiming 'Lorraine says No.'[11] The following month Creusot-Loire, the leader of the heavy engineering industry with a long and glorious past, which employed 300,000 workers, was abandoned by the Socialists and went bankrupt. Finally, the Renault car works at Billancourt, for so long the fortress of the Communist-dominated labour movement, closed in 1992, also under a Socialist government. It was the end of an era; but planned for three years with packages of retraining and retirement, and undertaken with the co-operation of the unions, it disappeared with more of a whimper than a bang.

The end of the working class

Now not only the traditional working class—based in heavy industry, working-class communities, and characterized by militant politics—but also the working class as a whole became an endangered species. The proportion of the working population employed in industry reached a peak in 1975 of 39 per cent, then fell to 34 per cent in 1982 and 29 per cent in 1992. Meanwhile the proportion employed in the tertiary sector grew from 57 per cent in 1982 to 65 per cent in 1992. Looked at another way, the number of industrial workers declined by 10.5 per cent from 8.1 million in 1975 to 7.25 million in 1989. As industry was modernized by automation and computerization, workers' skills became outdated or unwanted. The process of erosion affected unskilled workers in large-scale industry first, then unskilled workers in small businesses, then skilled workers in large businesses. The only branch that held up were skilled workers in small businesses, especially repair and maintenance firms, which served as a bridge between industry and the expanding tertiary sector. Whereas in the heyday of the proletariat there was a bunching of semi-skilled workers in the centre, marginalizing skilled artisans and unskilled casual workers, now a line divided the working class through the middle. At the top end of the scale workers were required to have computer skills, and to oversee machinery rather than operate it, much more like white-collar workers. Whereas in 1975 foremen had been designated 'workers' by INSEE, in 1982 they were moved into the 'intermediate professions' bracket along with technicians and middle management. At the bottom end of the scale was a growing mass of those who had no skills, or whose skills were obsolete, and whose employment was increasingly precarious.[12]

The fragmentation of the working class was demonstrated also by the decline of trade unionism and working-class consciousness. It has been calculated that trade unions lost about two-thirds of their members between 1974 and 1993, while in 1988 only about 9 per cent of the work-force was unionized. The decline may be explained by the erosion of the manufacturing base of the labour movement, the discrediting of the CGT by association with the PCF, and of the CFDT by association with the Socialist government. It was also a response to the failure of collective action to stop the decline of industry and displacement of collective goals by individual aspirations to property-ownership and family life. A survey carried out in 1984–5 showed among workers at Renault Industrial Vehicles in Caen a marked difference between those born before 1937 and those born after 1950. Whereas the older workers were usually of rural origin, ill-educated, had been the shock-troops of Taylorism, and joined but then left a union, the younger workers were more urban, having grown up in Caen, and were educated far above the monotonous requirements of assembly-line production. They had never joined a union but counted on individual success, bought houses in the outer suburbs of Caen in the later 1970s, had few children and desired a successful career for them, wanting to give them 'a good situation' rather than 'a trade'.[13] They had no working-class consciousness, and had probably ceased to regard themselves as working class.

A rural France without peasants

One of the great economic success stories of France was its agribusiness. From being a country of small, not very efficient, and highly protected farmers, France became the foremost agricultural nation in the EEC, accounting for 26 per cent of the Community's agricultural produce in 1985 (21 per cent provided by Italy, 17 per cent by West Germany, and 12 per cent by the UK) and the second largest exporter of farm produce in the world after the USA. The main reason, however, for this burst of production was that overproduction was no longer a concern of farmers. Whatever they produced, the CAP guaranteed the price, imposed Community preference against cheaper outside produce, and subsidized exports from the Community. The only grievances farmers could have were that the guaranteed price was not high enough, or that competition was being increased by the entry of new member states such as Greece in 1981 and Spain and Portugal in 1986.

The result was that by the 1970s the EEC was overwhelmed by wine lakes and butter mountains, most of the latter being sold to the Soviet Union in 1973 at a knock-down price. At the same time small farms were still being cushioned: the average dairy farm in Great Britain had 40 cows, that in the Netherlands 24, but that in France only 10. Even with price support such farms were uneconomic, so that French farming families were obliged to supplement their farm income with income earned outside farming. In 1980 two-thirds of farms had at least one outside income, and 21 per cent of farmers themselves also worked part-time outside.

As early as 1968 the EEC saw this problem, and the Mansholt Plan of that year recommended reducing the number of farmers in the Community by 50 per cent and the amount of agricultural land by 7 per cent. Larger, more modern, more efficient farms producing more cheaply would become the norm. This plan was adopted by European ministers in 1971 but not applied to France until 1976, because of the opposition of agricultural unions. The plan, in fact, hardly threatened the larger and more successful farmers who dominated the big unions, the FNSEA and the CNJA, but it could be used as a weapon against the smaller and less successful farmers. These were organized in the Mouvement de Défense des Exploitants Familiaux, which mouthed Marxist slogans and denounced the big unions, the Association Nation-ale des Paysans-Travailleurs, which broke way from the CNJA in 1974, and the Fédération Nationale des Syndicats Paysans, which broke away from the FNSEA in February 1982. The discontented unions received the backing of Mitterrand's agriculture minister, Édith Cresson, much to the fury of the FNSEA, which organized a march of 100,000 farmers through the centre of Paris on 23 March 1982 to reassert its hegemony, demanding a 16 per cent increase in farm prices and shouting 'Cresson, démission!' ('Cresson, resign!').[14]

The protests of the smaller unions would be ignored, not least by Michel Rocard, who replaced Cresson as Agriculture Minister in 1983. The EEC pressed ahead with its plans. In 1984 it imposed milk quotas which set small milk farmers against large ones in the west of France and drove 23 per cent of them off the land over the following year. In 1988 a fifth of all arable land was required to be taken out of cultivation as 'set-aside', and not all could be put to use as golf courses, theme parks, or lakes for wind-surfers. In 1992 the European Commission decided on a fundamental reform of the CAP, involving the progressive lowering of guaranteed prices. Opposing this measure, 200,000 farmers, organized by the FNSEA, demonstrated in Paris on 29 October 1991. However, the EEC's strategy to reduce subsidies, cut production, and leave only the most efficient on the land had the desired effect. The proportion of the working population involved in agriculture, which was 20 per cent in 1962, fell to under 10 per cent in 1975 and 5 per cent in 1990. The number of farmers fell from 1.6 million in 1975 to 1 million in 1990, and estimates of the number of farmers France actually needed ranged from 250,000, who were economically necessary, to 500,000, which might be socially acceptable.[15] About half the farms in the period 1979–85 were taken over by descendants, but only 28 per cent of farmers' sons aged between 25 and 39 became farmers, so that purchase by outsiders and growing concentration was the dominant pattern. Farmers in 1990 were essentially successful businessmen, well educated and indeed no longer rural beings: three-quarters of them lived either in a town or just next to one.

Paradoxically, the census of 1982 revealed that for the first time in a century, the rural population had grown more rapidly in the period

1975–82 than the urban population. The census of 1990 confirmed this pattern. It did not affect the countryside of 'la France profonde' around the Alps, Pyrenees, and Massif Central, where the rural population continued to decline. The purchase of holiday homes by Germans in Catalonia, by Dutch and Belgians in the Cévennes, and by British in the Dordogne only underlined the isolation of the local inhabitants. The growth areas were within a radius of 20–30 km. outside provincial towns, and 30–70 km. outside Paris. While the inner cities were being lovingly restored and rejuvenated by young professionals, the outer suburbs of high-rise estates, built in the 1960s, totally lacking in amenities and jobs, were becoming degraded and filling up with immigrants. Les Minguettes outside Lyon, with 36,000 residents and 40 per cent unemployment among young immigrants, exploded in the summer of 1981 in a frenzy of car-burning, joy-riding, and shootings. To escape this, city-dwellers with young families were moving into rural districts, promised 'happiness within everyone's grasp' by estate agents, close to nature, snug in their custom-built bungalow with garden, garage, and barbecue, yet within commuting distance of towns and serviced by out-of-town supermarkets. The process gave rise to a new term, 'rurbanization', and to new country-dwellers, who had nothing in common with the farming communities they were displacing.

The decline of family life, immigrants, and the new poor

The population of France increased very slowly between 1975 and 1990, from 52.6 to 56.6 million. Death rates decreased and life expectancy rose between 1970 and 1990 from 69 to 73 for men and from 77 to 81 for women. The main reason for the stagnant population was the birth rate, which plummeted after 1965. From 18.1 per 1,000 in 1964 it fell to 13.4 per 1,000 in 1990. Whereas the average number of children per family was three in 1964, it was two in 1975 and 1.8 in 1986. This was explained less by the return of women to work in greater numbers, since the birth rate for working and non-working mothers was roughly the same, than by a reaction against the previous generation of large families and the demand of women to have greater equality and control over their own lives.

This trend was reflected also in marriage rates, which almost halved in France between 1965 and 1985, and were lower than those in Great Britain, West Germany, and Denmark, though higher than those in

Sweden. The number of households of single people increased rapidly after 1975, from 2.8 million in 1962 to 4.8 million in 1982. In Paris one person in two lived alone in 1982, two-thirds of them women. Cohabitation outside marriage increased from 4 per cent of couples in 1975 to 7 per cent in 1985, higher than in Great Britain and the USA (though much less than in Sweden), and the percentage of children born outside marriage grew from 8.5 per cent in 1975 to 18.0 per cent in 1984, about the same as in Great Britain and the USA. At the same time the divorce rate shot up, from 12 per cent of marriages in 1970 to 22 per cent in 1980 and 32 per cent in 1990. Nearly two-thirds of these were demanded by women in 1973, nearly three-quarters of them in 1985.

Though much fuss was made of immigration, foreigners made up a smaller proportion of population in France in 1990 (6.4 per cent) than they had in 1931 (6.6 per cent). New immigration was stopped by the government in 1974, and the only entries allowed were to reunite families where (in general) the male partner had gone ahead alone to find work. That said, fears were expressed about illegal immigrants who did not figure in the statistics, and immigrants also tended to be increasingly visible. Of the total immigrant population, the proportion from North Africa (Algeria, Morocco, Tunisia) rose from 32.3 per cent in 1975 to 39.2 in 1990, while Turks increased from 0.4 per cent to 5.6 per cent of the total. The immigrant population was heavily urban and Parisian. Figures for 1982 showed that 58 per cent of them lived in the Paris region. One in six Parisians in 1982 was foreign (30 per cent of them from North Africa) and one in ten Marseillais (69 per cent from North Africa). One in five births in Paris was to a foreign mother, mainly Portuguese or North African. Where they could find work, immigrants concentrated in semi-skilled and unskilled jobs. Fifty per cent of Algerians were classified as unskilled in 1982, compared to 24 per cent of Italians and 13 per cent of the French, and 22 per cent of Algerians were unemployed compared to a national average of 8.6.

Another crucial development in France was the growth of inequality of income, a widening disparity between rich and poor. While the difference between the income of the richest 10 per cent of the population and the poorest 10 per cent fell in the period 1969–83, after 1983 the gap widened. In the first period, the erosion of income from property by inflation, the spread of salaried employment, the indexation of the SMIC to prices, and the increase in old-age pensions all served to reduce the difference between rich and poor. After 1983, however, and paradoxically under an extended period of Socialist government, the policy of austerity cut back the rise of low incomes, by increasing the tax and national insurance contributions of the low-paid, breaking the price-indexation of the SMIC, and expanding the 'reserve army' of unemployed.

Behind the bald statistic of the rate of unemployment in fact lay a much more complex situation. In 1986, for example, the unemployment rate was 10.4 per cent. But a more complete study argued that 11.7 million people or 47 per cent of the working population was at various degrees at risk of unemployment. Of these 6.8 million (27.2 per cent) had a stable job but thought they might lose it within two years. The remaining 4.9 million had difficulties finding a regular job or were actually unemployed. Within this group 1.8 million (7.4 per cent of the working population) were unemployed and increasingly out of touch with the labour market, 1.3 million were long-term unemployed, not having worked for two years, 850,000 were classed as marginal to society, and 250,000 were outside society altogether.[16]

Unemployment was increasingly the main factor behind poverty. No longer in the 1980s were the poor the aged: a lifetime of employment and generous old-age pensions saw to that. Neither were they exclusively the social marginals of the 'Fourth World'. The New Poor of the 1980s originated in the heart of society, but had dropped down below a given income level by reason of some catastrophe, be it unemploy-

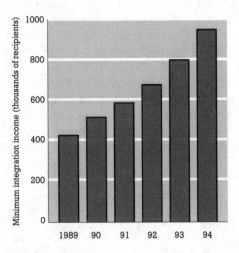

Fig. 6. Growth of Number of Recipients of Minimum Integration Income (RMI), 1989–1994.
Source: Le Monde, 13 December 1994.

ment, sickness, or divorce. They included unemployed architects and managers as much as bankrupt shopkeepers or workers laid off. Those who had been employed qualified for unemployment benefit; those who had not did not. A Revenu Minimum d'Insertion (Minimum Integration Income, or RMI) was invented in 1988 to cover such people; it made grants to 400,000 individuals in 1989, 800,000 in 1993, and nearly a million in 1994. Over half those funded were individuals, but single mothers were a rising proportion, 18 per cent in the first half of 1991, 22 per cent at the end.[17] Another fifth were reckoned to be drawn from the 250,000 excluded from society, but at best only 30 or 40 per cent of those 250,000 were covered, even by RMI. A survey conducted in Saint-Brieuc in 1991, indeed, drew a distinction between those receiving benefit intermittently, those receiving it permanently, often the inhabitants of degraded high-rise flats, and those who fell through the net altogether or never requested social services. Squatting or living in caravans on waste ground, doing odd jobs or seasonal work for farmers, they thrived on a certain roguish freedom.[18] They were nothing if not the 'Fourth World' reincarnate.

5

The One and Indivisible Republic?

There were four cardinal principles of the French Republic. The first was that education was universal, secular, compulsory, and free, providing both a uniform education for all future citizens and an equality of opportunity that opened careers to talent alone. The second was that all citizens were equal under the law, and that the law dealt equitably with all, whatever their class, race, or gender. The third was that France was a centralized, unitary state and One and Indivisible Republic, in which the laws made by a single legislature, articulating the will of the sovereign people, were applied equally in all parts of France. The fourth was that the French were not, like the Germans, a *Volk*, bound together by ties of blood, soil, and language, but a body of citizens who had come together at the French Revolution to establish a new social contract that would guarantee the rights of man and the citizen, found out by reason.

These principles of equality were the founding myths of the Republic. They were powerful and persuasive, defined the content and parameters of French political discourse, and made the questioning of those principles extremely difficult. In fact they disguised, and were intended to disguise, radical inequalities.

The first was the dominance of an élite which, though recruited meritocratically, in fact reflected social and economic inequalities and constituted, 200 years after the end of the Ancien Régime, a new privileged order. The second was the dominance of one gender over another, for, despite the belated granting of political rights to women at the Liberation, women still suffered the consequences of a social and political order based on the patriarchal family, the control of private property by married men, and the separation of public and private spheres. This had been strengthened rather than weakened by the French revolutionaries and Napoleon, and indeed it had been widely argued down to 1945 that, because women were held to be prey to the Catholic religion and thus anti-republicanism, to give women the vote would endanger the Republic itself. The third was the dominance of a political class of politicians and bureaucrats who controlled the levers of the centralized state and refused any idea of sharing power with local or regional interest groups. These partisans of a centralized Republic, who may be called Jacobins, did not hesitate to argue that localism and regionalism was the agenda of reactionaries and that to allow any decentralization in France would play into the hands of counter-revolution. The fourth, in spite of the idea of the French nation as a body of citizens who wished to be bound by the terms of a new social contract, was the dominance of those who had been born on French soil, of French parentage, and who had assimilated French enlightenment and civilization over those who had not. Thus, though in theory it was possible to acquire French nationality by an act of will, in practice the French refused to consider that anyone could be properly French if they were foreign in origin, spoke a different language, or threatened the idea of the secular state by demanding public recognition of their religion. This made the integration of immigrants into French society both extremely problematic from the ideological point of view and extremely painful in practice. The One and Indivisible Republic and a multicultural and multireligious society appeared to be difficult to reconcile.

Democracy and inequality in education

In 1986 Antoine Prost, one of the greatest authorities on modern French education, published a book entitled *Has French Education Become Democratized?*[1] His answer was generally positive. Demand for education increased dramatically between 1950 and 1975 as real wages rose, supported by family allowances and social security, the economy boomed and families sought a better future for their children. At the same time the government needed to develop education in order to provide a trained work-force for the modernizing economy. Thus national education's share of the total budget doubled between 1956 and 1965, 2,354 colleges were built between 1966 and 1975, and the school population increased from 6 million to 13 million between 1959 and 1980.

More significantly, the traditional barrier between primary education and secondary education, one for the bourgeoisie and the other for the masses, was gradually broken down. State primary education was fully free after 1881, and this became the rule in state secondary education from 1927, although whether or not a child went into a secondary school at the age of 12 and began Latin with a view to the *baccalauréat* at 18, or stayed within the primary and higher primary system to obtain the *certificat d'études* at 14, remained a decisive issue. Then in 1959 the school-leaving age was raised from 14 to 16. In 1958 30 per cent of pupils left at the age of 14. In 1962 32 per cent of boys and 36 per cent of girls stayed at school until the age of 18, and in 1987 those figures rose to over 70 per cent of boys and nearly 80 per cent of girls.

The raising of the school-leaving age made the coexistence of two different systems impossible, however, and a reform of 1962 established a common secondary education, in *collèges d'enseignement secondaire*, from 11 to 16. Teachers and pupils aged 11 to 14, who had hitherto been accommodated in the higher primary system, were moved bodily into the colleges. Latin, which for so long had defined secondary, bourgeois education, was abolished in the first year of secondary education (*sixième*) in 1968. Divisions persisted between 'long' classical and modern streams, 'short' modern streams and 'transition' (12–14), and practical' (14–16) streams, which reflected the divisions between former secondary and former primary teachers and pupils; but the reform of René Haby in 1975 abolished the 'transition' classes and established a common curriculum up to the age of 14. The proportion of French children with the *baccalauréat*, taken at 18, increased from 8 per cent of boys and 7 per cent of girls in 1962 to 23 per cent of boys

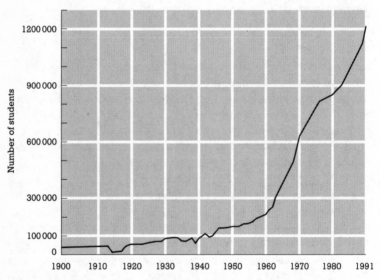

FIG. 7. Growth of Number of University Students, 1990–1991.
Source: L'Etat de la France 94–95. (Paris, Éditions La Découverte, 1994).

and 28 per cent of girls in 1987. This proportion was exceeded only by the USA, Canada, West Germany, and Italy. In 1984 the minister of national education laid down the challenge of 80 per cent of French children with the *baccalauréat* by the year 2000, and in 1991 the result, including the technical *baccalauréat*, was already 58 per cent. Meanwhile the university population expanded from 118,000 in 1945–6 and 244,000 in 1961–2 to 800,000 in the late 1970s and 1,200,000 in 1990.

Whether this expansion and democratization meant equality of opportunity and open competition for jobs was another matter. It was countered by constant pressure from families to limit competition, rig the market, and permit the blatant transmission of wealth, status, and privilege behind the façade of meritocracy. It is not simply that the education system was a tool of the dominant classes and used by them to 'reproduce' the existing social structure, which in any case was changing fast. The education system was not a ladder but a series of hurdles. As the hurdles became higher, each social group had to measure the costs and benefits of the next lap for itself, and each made different decisions. By and large the lower classes opted for employment at an earlier stage, and the middle classes continued to invest in the later stages. Moreover, whereas the educational level reached by a child's parents was the main factor determining how far a child would go in the early stages, such as whether he or she would go into secondary education in the period before the reform of 1962, in the later stages, and in respect of the kind of job ultimately obtained, the main determining factor was the social background of the parent. A survey of 1970 thus demonstrated that the chances of becoming senior managers for those who had the *baccalauréat* was 25 per cent for sons of industrial workers or white-collar workers, 30 per cent for sons of artisans and small shopkeepers, 39 per cent for sons of middle managers, and 46 per cent of sons of senior managers.[2]

The importance of social background in determining educational outcome has been confirmed by other research. French students who fail to meet the required educational attainment were required to repeat a year, which had a knock-on effect because of the imposition of age limits formally or informally at later stages. A survey of 1979–80 showed that, of those required to repeat the first primary year (aged 6), only 5 per cent had parents who were professionals or senior managers, while 20 per cent had parents who were semi-skilled or unskilled workers or without employment.[3] The *collège* became a battleground on which crucial decisions about which stream pupils would be allocated to were taken. The 'long' classical or modern stream held out for the middle classes the prospect of a fast track to a *baccalauréat* at 16, leaving two years to prepare for the *grandes écoles*; the 'short' modern stream gave middle-class children a chance to catch up and a chance of promotion for working-class children; while the practical classes (called *classes pré-professionnelles de niveau* after 1975) led to the *certificat d'aptitude professionelle* (CAP) for future skilled workers or to the *brevet d'enseignement professionnel* (BEP) for future small employers. Many working-class pupils who would have done well in the higher primary system reacted against a third-class treatment in the colleges and failed to progress in the lycées which educated children over the age of 16. In the 1980s a *baccalauréat technique* was introduced, taught in technical lycées, but it was in no sense designed to lead to the *grandes écoles*. A study of pupils who reached the top classes of lycées (including technical lycées) in 1952, 1967, and 1980 showed that the proportion of working-class children expanded from 10 per cent in 1952 to 18 per cent in 1967, but (notwithstanding the slight decline of the working class) was still 18 per cent in 1980.[4] Meanwhile the proportion of children from liberal professional or senior management backgrounds fell from 28 to 18 per cent between 1967 and 1980, but rose sharply again to 30 per cent in 1980.

Whereas since the nineteenth century the possession of the *baccalauréat* had been the dividing-line between the bourgeoisie and everyone else, the increase in the number of pupils with the *baccalauréat* simply devalued the qualification and meant that the selection process had to take place at a later point. Since the *baccalauréat* gave an automatic right of entry to university, examinations at the end of the first university year became the main point for thinning the ranks. It also gave rise to a two-tier system of higher education, broadly between the 900,000 students in the universities, which were non-selective, and offered no guarantee of employment, and the 300,000 in the *grandes écoles* and Instituts Universitaires de Technologie (IUT), founded in 1966–7, which were selective and in general guaranteed jobs. Another distinction, made by the sociologist Pierre Bourdieu, was between the *grande porte* and the *petite porte*. The *grande porte* was that of the élite of the *grandes écoles*, such as the École Polytechnique, the École Nationale d'Administration (ENA), the École Normale Supérieure, the École Centrale, the École des Hautes Études Commerciales (HEC), and the École Supérieure des Sciences Économiques et Commerciales (ESSEC). The *petite porte* that of the universities, the IUT, and a mass of smaller specialized schools. The first trained generalists for the top posts in the civil service, politics, industry, commerce, and research, the second trained specialists for the so-called 'intermediary professions': technicians, middle managers, primary and secondary school teachers.[5] The distinction became even more pronounced after the student revolution of 1968. The *grandes écoles* were exempted from the subsequent university reform, and President Pompidou, opening new buildings for the École Centrale in 1969 said, 'at a moment when our university is profoundly shaken and seeks feverishly to find its own equilibrium, our *grandes écoles* remain the most solid bastions for the preparation of the nation's leaders.'[6] Thus the *grande porte* was increasingly monopolized by the social élite and the old-boy network, while the latter recruited lower and lower down the social scale and increasingly among women. Non-academic children of the social élite developed a network of their own schools, such as the Institut Européen d'Administration des Affaires (INSEAD), founded in 1958, and the European Business School, which required some work experience and saved them from the state universities.

Social mobility and the French élite

There is no doubt that the period of rapid economic change in the thirty years after the war saw an increase in social mobility. A good deal of this was explained simply by changes in the structure of the working population, but it is also clear that French society became more flexible and that chances of social promotion or demotion increased. A survey measuring the social position that sons had reached by the age of 40–59, measured in 1953 and again in 1977, showed that some social categories were not reproducing themselves from one generation to the next. While the working class was fairly stable, 62 per cent of sons of workers becoming workers in 1953 and 58 per cent still becoming workers in 1977, the percentage of farmers' sons who became farmers fell from 60 to 38, that of sons of artisans and small shopkeepers who followed their fathers declined from 48 to 22, and that of sons of businessmen and those in the liberal professions who in turn became businessmen or entered the liberal professions declined from 43 to 21. The percentage of sons who became senior managers (including *ingénieurs* or teachers in secondary or higher education) rose between 1953 and 1977 from 3 to 12 among artisans and small shopkeepers, and from 12 to 26 among businessmen and the liberal professions, but the percentage that sank into the working class rose from 26 to 32 among artisans and small shopkeepers and from 17 to 35 per cent among farmers. There was an increase in career mobility as well as in inter-generational mobility. Of the first generation of artisans and small shopkeepers, born in 1894–1913 and measured in 1953, three-quarters had remained in the same occupation all their lives, but of the second generation, born in 1911–30 and measured in 1977, only a third had.[7]

Such pictures of social mobility must be qualified in two respects. First, society became less mobile in the period after 1973. Second, some categories of society were more inclined to social mobility than others. A survey tracing the destinies of sons aged 40–59 in 1977 and 1985 established that the percentage of employers' sons who became employers declined marginally from 30 to 29 and that of workers sons' who became workers fell marginally from 52 to 49 between 1977 and 1985, while that of farmers' sons who became farmers remained constant at 33. At the bottom of society, farmers and workers were something of a caste, while at the top, exit from the dominant class was rare. The percentage of sons of senior managers and allied professions who remained in the same bracket rose from 58 per cent in 1977 to 60 per cent in 1985. In the middle of society there was much less definition between the different categories and, by dint of being in the middle, access to other categories above and below was easier. Thus, taking the

TABLE 4. *Patterns of Social Mobility, 1953–1977*

Fathers		Sons (aged 40–59)						
		(1)	(2)	(3)	(4)	(5)	(6)	(7)
(1) Farmers	1953	**60**	8	1	2	2	5	17
	1977	**38**	6	2	3	4	9	35
(2) Artisans, shopkeepers	1953	4	**48**	3	3	6	8	26
	1977	3	**22**	6	12	10	14	32
(3) Liberal Professions, businessmen	1953	3	10	**43**	12	7	7	17
	1977	1	14	**21**	26	12	10	16
(4) Senior management	1953	4	16	6	**41**	17	14	2
	1977	2	4	10	**43**	22	9	10
(5) Middle management	1953	0	16	12	16	**16**	19	21
	1977	1	4	6	30	**30**	11	18
(6) White-collar	1953	5	10	4	12	14	**17**	38
	1977	1	7	2	16	21	**17**	35
(7) Workers	1953	3	11	1	1	9	11	**62**
	1977	1	7	2	6	11	14	**58**

Source: Claud Thélot, *Tel père, tel fils?* (Paris, Dunod, 1982), 46.

two years 1977 and 1985 together, only 14–15 per cent of sons of
white-collar workers followed their fathers; a tenth became employers,
a fifth became senior managers, a fifth workers, and a third joined
middle management. Of the sons of those in middle management
positions, a third followed their fathers, a third became senior managers, a sixth workers, and a tenth each employers and white-collar
workers. These patterns of mobility were confirmed by patterns of
intermarriage. At the extremes of society in 1985, about 30 per cent of
sons and daughters of senior managers and the liberal professions
married daughters and sons of the same, while over 50 per cent of sons
and daughters of workers and farmers respectively married daughters
and sons of workers and farmers. The equivalent figures, however, for
the sons and daughters of white-collar workers was 15 per cent and of
those in middle management a mere 14 per cent.[8]

At the very summit of French society, the picture was not one of
social mobility but of the interlocking and mutual reinforcement of
social background, private wealth, graduation from the élite schools,
and the monopolization of economic, administrative, and political
power. Although the emergence of a new middle class of *cadres*
reflected the growing weight of salaried income relative to family
fortune or *patrimoine* within the bourgeoisie, immense family fortunes

TABLE 5. *Patterns of Social Mobility, 1977–1985*

Fathers			Sons (aged 40–59)					
			(1)	(2)	(3)	(4)	(5)	(6)
(1)	Farmers	1977	**33**	10	5	11	7	34
		1985	**34**	9	5	12	7	34
(2)	Employers	1977	2	**30**	20	21	8	20
		1985	2	**29**	20	19	7	23
(3)	Senior	1977	1	10	**58**	22	4	5
	management	1985	0	9	**60**	21	6	4
(4)	Middle	1977	0	9	31	**36**	10	14
	management	1985	0	10	32	**31**	9	18
(5)	White-	1977	1	9	22	31	**15**	23
	collar	1985	0	10	23	32	**14**	22
(6)	Workers	1977	1	9	7	21	10	**52**
		1985	1	10	8	22	10	**48**

Source: Dominique Merllié and Jean Prévot, *La Mobilité sociale* (Paris, La Découverte, 1991), 62.

were still found in French society. A minority were based on landed estates of over 500 hectares, but most were based on industrial and banking wealth going back to the nineteenth century or more recent fortunes derived from international finance capital. Confronted by the Impôt sur les Grandes Fortunes of 1981, over 100,000 taxpayers declared a fortune of over 3 million francs, 141 of over 100 million francs, and five of over a billion. This was probably an underestimate, given an estimate of 1977 of fifteen fortunes of over a billion.[9]

In theory the *grandes écoles* were meritocratic but in practice, as we have seen, they were dominated by the social élite. The *grandes écoles* fed into the *grands corps d'État*, the great offices of state at the apex of the bureaucracy. The top graduates from Polytechnique went on to the Écoles des Mines or Ponts et Chaussées and join the Corps des Mines or Corps des Ponts et Chaussées. Those who passed out top from the École Nationale d'Administration, which recruited indirectly from other *grandes écoles*, including the Polytechnique, colonized the Inspection des Finances, the Cour des Comptes, and the Conseil d'État. The *grands corps* were the powerhouses of the state, and established a tentacular grasp on other key institutions. After ten or fifteen years in one of the *grands corps*, members of this power group became directors or chairmen of the large banks or industries linked to the state by being nationalized, semi-public, or having the state as a major customer. These sideways moves into industry, known commonly as *pantouflage*, allowed the administrative élite the opportunity to make more money and permitted industry important contacts with government. There was a world of difference between employers of private, often family businesses, embedded in the regions and concerned only with balance sheets, and the *patrons d'État*, whose families belonged to the high civil service or liberal professions rather than to industry, who ran the merchant banks and energy and transport industries, and who were well educated, well connected, and close to power.

A route out of the *grands corps* that was more directly linked to power was into the ministerial *cabinets*, the private offices of ministers. They were generally composed of a director, head of secretariat, and seven other counsellors. Ninety per cent of them were seconded from the civil service under the Fifth Republic (against 60 per cent in the Third), three-quarters of them from the *grands corps*; the director was usually drawn from the Conseil d'État. A spell of three, six, or ten years in a ministerial office was an excellent jumping-off point for a number of careers. Becoming a chairman of a large public or semi-public company was always an option, as was a return to the civil service, especially as the top posts were in the gift of ministers. Above all, though, it was the grounding of a political career. A start could be made as *suppléant*, elected alongside deputies under the Fifth Republic and taking over as deputy should the elected deputy be made a minister. Moreover, among Gaullists it was more common to become a deputy by being 'parachuted' into an unknown seat than by working a passage up from municipal councillor, mayor, and *conseiller-général*, so that members of ministerial cabinets were well placed for this. Thus Jacques Chirac, a graduate of the ENA, began his career in the Cour des Comptes and served in the private office of Georges Pompidou, before being elected in the Corrèze in 1967, thanks to the patronage of the department's grand old man, Henri Queuille, and becoming a junior minister in the government of Couve de Murville in 1968. In this way members of the *grands corps* constituted a new *noblesse de robe*, from which were bred the *ministrables* and ministers of the Fifth Republic.

The Liberation and women

At the Liberation, French women were liberated as French citizens rather than as women. As women, they enjoyed the Liberation as a never-ending party, going with American officers and GIs and being given nylons and chewing gum. Even before then, however, an ordinance of the Free French authorities of 21 April 1944 established that women as well as men would be entitled to elect the Constituent Assembly of the new Republic, and women in fact voted for the first time in France in the municipal elections of 29 April 1945. Symbolically, this was recognition of the part that women had played in the Resistance; realistically, since women had been denied the vote under the Third Republic on the grounds that they would vote for royalists as instructed by their parish priests, it was a ploy by de Gaulle to offset the landslide it was feared the Communists would obtain.

Granting women the vote made only a limited difference to political life. Women had not seen Resistance to the Germans as politics, and at the Liberation male party politicians were anxious to recover the influence they had been denied since 1940. Under the system of proportional representation some women were included on electoral lists, but not at or near the top. In the elections to the first National Assembly in November 1946 women thus provided 14 per cent of the candidates but less than 7 per cent of the deputies. This result, however—42 women deputies—was an improvement on the 33 women elected to the first Constituent Assembly and 30 elected to the second, and would never be matched in a National Assembly election subsequently. De Gaulle's gamble paid off. In 1946 and 1951 fewer women than men voted for the PCF, SFIO, and Radicals; more women than men voted for the MRP, Moderates, or RPR, the parties of the Right or at least (given the difficulty of classifying the MRP) parties which had traditional views on the family. When women stood up to debate in parliament, it was never to express an opinion on the very public sphere of foreign policy or defence, nor, for some reason, on agriculture, but usually on matters that were related to the private sphere, such as food supply, the family, population, housing, health, and education. Similarly, the first woman minister in France (bearing in mind that the three women appointed to the government in 1936 had been under-secretaries of state), Germaine Poinso-Chapuis, a lawyer, campaigner against prostitution and alcoholism, member of the MRP, and vice-president of the municipal council of Marseille in 1945, was made minister of health in the Schuman government of 1947–48.

After 1950 women were excluded even from this well-defined women's sphere. Only 22 women deputies were returned in the elections of 1951, and 19 in 1956. With the Fifth Republic and the reintroduction of single-member constituencies, women were no longer favoured as candidates and their number went down to 8 in the National Assemblies of 1958 and 1962, 10 in 1967, 8 in the Gaullist Parliament of 1968, fewer than 2 per cent of the deputies. This did not prevent women from fervently supporting Charles de Gaulle, on the contrary. A poll of 1962 revealed that 64 per cent of them were satisfied with his leadership and only 18 per cent dissatisfied, while the equivalent figures for men were 57 and 29 per cent.[10] Similarly, 62 per cent of women voted for de Gaulle in the first ballot of the presidential elections in 1965 and only 38 per cent for Mitterrand, even though (or because) Mitterrand had pronounced in favour of freely available birth control. Yet de Gaulle, once asked about creating a minister of women's affairs, retorted, 'A ministry? why not an under-secretaryship of state for knitting?'[11]

A new look for women

It may be assumed that women were politically uneducated or even duped in this period. The truth is, however, that women's priorities were very specific at the end of the war. Family life was privileged because families had been dislocated during the Occupation. Perhaps a total of 5 million individuals had been prisoners of war, deported, sent to Germany as forced labour, or expelled from their homes. The Liberation was a moment of family reunion, or alternatively the recognition of family breakdown. Men returned to work, women retired to the home to have children. After years of deprivation, building a comfortable home life and raising one's standard of living became all-consuming tasks. New domestic appliances such as washing-machines and fridges were now available to make housework easier: if women were liberated by anything in the 1950s it was by Moulinex and Bendix.

Again, after the hardship and shortages of the Occupation, women wanted to become feminine once again. *Elle*, which appeared in November 1945, edited by Françoise Giroud, sold 110,000 copies when it started, 500,000 in 1950, and a million copies in the 1960s. It advertised washing machines, electric irons, and Pyrex casseroles. It taught women how to be seductive and to keep their men. The refeminization of woman was proclaimed in February 1947 when Christian Dior launched his New Look. The 'soldier-woman' with square shoulders and culottes under an overcoat gave way to the 'flower-woman' with rounded shoulders, a heightened bust, narrow waist, and immense fan-like skirt 14 metres in circumference. And women escaped into fairy-tale land when the 26-year-old American actress Grace Kelly met her Prince Charming at the Cannes Film Festival in 1955 and became Princess of Monaco.

Into this atmosphere Simone de Beauvoir's *The Second Sex* plunged like a lead balloon. Published in 1949, it argued that women were now equal to men as far as abstract civil and political rights were concerned, but that so long as women were economically dependent on men equality was merely an illusion. There was inequality within the marriage, as the husband was still legally head of the household, and had the right to choose where the family lived to suit his job, while the woman could not divorce without financial suffering. A few women did have successful careers, but they found that they had to deny themselves family life, even their femininity, or have them denied to them. Last but not least, women were enslaved by the law of 1920 that criminalized abortion and prohibited the dissemination of contraceptive devices or information about them. De Beauvoir asserted that there were nearly a million abortions a year in France, as many as live births, but only working-class women were in danger of punishment, or death following back-street abortions, because middle-class women could obtain 'therapeutic' abortions in private clinics or pay for treatment in Switzerland.

Simone de Beauvoir's book was little read or understood at the time. She had no contact with women's organizations, which tended to be bourgeois, concerned with questions of women's property, defended the professional interests of the minority of women who were successful, or espoused traditional views of the role of wives and mothers. Her ideas subsequently found an echo in the Mouvement Démocratique Féminin (Women's Democratic Movement, or MDF), founded in 1964 by Marie-Thérèse Eyquem, Colette Audry, and Yvette Roudy. This was a think-tank and club in the orbit of the Convention des Institutions Républicaines of François Mitterrand, who was converted to their ideas on birth control. But pressure groups had to present their demands in ways least likely to provoke controversy. An organization set up in 1956 to campaign to liberalise the law on contraception called itself the Association Maternité Heureuse (Happy Motherhood Association) and argued its case from the standpoint of women's health and happy motherhood (not exhausted by endless pregnancies) as a precondition of a happy family, not from that of sexual liberation. In 1960 it changed its name to the Mouvement Français pour le Planning Familial (French Family Planning Movement) and opened centres in Grenoble and Paris. It secured the support of some doctors and politicians, although it was fiercely opposed by other doctors, the Catholic Church, and the Communist party, which regarded contraception as a bourgeois vice and calculated to restrict the size of the revolutionary proletarian army. A bill sponsored by the Gaullist deputy Lucien Neuwirth, who had discovered the evils of unwanted children while deputy-mayor of Saint-Étienne, sponsored a bill that became law in December 1967. It was only a small step forward, allowing contraception on prescription, but still limited advertisement, required parental authority for minors, and offered no reimbursement. The Catholic Church promptly re-affirmed its opposition to contraception in the 1968 encyclical *Humanae Vitae*, and in 1972 only 6 per cent of French women said that they used the pill.

Women and liberation

The feminist movement proper in France emerged from the cauldron of May 1968. The revolution demanded free love as part of the agenda of total liberation, outside marriage, and for pleasure rather than procreation. Women who had been involved briefly in the MDF gained experience in brain-storming sessions and political activism inde-pendent of political parties, and found a new language to challenge patriarchal oppression and conventional views on marriage, the family, and sexuality. In May 1970, at a meeting in Vincennes university, they broke with male activists who called them 'mal baisées'. That November they disrupted an Estates-General called by *Elle* magazine to Versailles, on the grounds that it locked women into traditional roles, and laid a wreath at the Tomb of the Unknown Warrior at the Arc de Triomphe inscribed, 'There is someone more unknown than the Unknown Warrior: his wife'.[12] *Le Torchon brûle*, an ephemeral magazine that gave an angry and passionate voice to feminists, also appeared in 1970. In 1973 a group called *Psychanalyse et politique* (or *Psych et Po*) was formed. This took the view that the oppression of women was essentially psychological, the result of internalized repression, and that the main challenge for a woman was to 'chase the phallus from her head'.[13] These expressions of militancy and protest, which turned the weapon of ridicule and symbolic violence against their opponents, was called the Mouvement de Libération des Femmes (Women's Liberation Movement, or MLF) by the media, some time before the women appropriated it for themselves.

The next bombshell of the feminist movement was a manifesto demanding legal and free abortion, published in *Le Nouvel Observateur* on 5 April 1971 and for which militants had secured the names of 343 women who confessed to having had abortions, including Simone de Beauvoir, Françoise Sagan, Marguerite Duras, and Catherine Deneuve. While the radical wing of the movement took to the streets on 20 November 1971 to demand abortion, singing 'Travail, Famille, Patrie, y en a marre' ('Work, Family, Fatherland, we've had enough of them'), Gisèle Halimi, the laywer of the 343, who had herself undergone an abortion for fear that her Tunisian father would kill her, founded an organization called Choisir which was the legal and respectable wing of the movement. Her moment of triumph was her successful defence in the Paris suburb of Bobigny of a schoolgirl charged with having had an abortion. The much-publicized verdict destroyed the law of 1920, and required new legislation on abortion. This was drafted by Simone Veil, the health minister of incoming president Giscard d'Estaing. Though it was a compromise, refusing reimbursement and for only a trial period of five years, it was passed with the support of Socialists and—after a change of heart—by the Communists, and promulgated on 17 January 1975.

If one central issue of women's rights was the right to contraception and abortion, a second was the right to equal treatment at work. Women had taken themselves out of the labour market after the war. Only 35 per cent of the working population were women in 1954, and the figure remained the same in 1968. Only after a law passed in 1965 was a wife given full control of her personal property and earnings, and even then the husband could veto her return to work in the interests of the family. Women began to return to work in the 1970s, and 45 per cent of the working population was female by 1985. A survey of 1973 showed that the new generation of women stopped work to raise a family for much less time than previous generations, 60 per cent for less than five years.[14] Work in itself did not, of course, confer liberation. Most women were trapped in low-paid jobs in sectors of the economy traditionally reserved for women such as the food, clothing, and electronics industries, retailing, and office work. In 1973 they accounted for 96 per cent of typists, 88 per cent of receptionists, 78 per cent of cashiers, and over 70 per cent of primary school teachers, social workers, and nurses. In shops, banks, and post offices the counter staff were overwhelmingly women and the supervisors invariably men. Only 10 per cent of managers but 72 per cent of those on the minimum wage were women.[15] The law of 1972 on equal wages for equal work remained a dead letter.

At the top end of the market some women were undeniably doing well. In 1978 the concept of the *nouvelle femme* was launched by *F-Magazine*. *F-Magazine* was founded that year by Claude Servan-Schreiber, wife of the Radical politician, and the writer Benoîte Groult. It attracted a pleiade of talented young journalists including Anne Sinclair and Christine Ockrent. Forswearing articles on beauty, fashion, or cooking, it held out a mirror to successful women who had found a happy balance between their careers and family life, and had achieved equality while losing none of their femininity. A survey carried out by the magazine amongst its readers in April 1978 found that two-thirds were married, but that 43 per cent had completed higher education, 55 per cent had resumed full-time work after having a family, and that the husbands of three-quarters of them helped in the home. Only 8 per cent of the women never wore makeup, 73 per cent had satisfactory or very satisfactory sex lives, 80 per cent thought that contraception was at the origin of recent changes in the condition of women, and 93 per cent that the legalization of abortion was progress. However, these liberated women saw themselves as external to the movement for women's liberation. Only 13 per cent were totally in favour of the movement and would take part if they could; 48 were sympathetic, 32 per cent did not like the means they used, and 3 per cent thought them 'ridiculous'. One respondent said bluntly that they were 'problem women'. The *nouvelle femme* thus evolved a 'soft' feminism which did not call into question the importance of feminity and, while benefiting from the struggles of 'hard' feminism, were careful to keep their distance from it.[16]

A third major problem for women was how to influence political decisions and to achieve power themselves. Simone Veil, who had the qualifications to be minister of justice, was given the traditionally female post of minister of health by Giscard d'Estaing in 1974. Giscard also invented a secretaryship of state for women's affairs, but the women who filled it were far from being feminists. The first, Françoise Giroud, had been editor of *Elle* before moving to *L'Express*; the second, Monique Pelletier, a lawyer and mother of seven, attacked feminism by arguing that 'no form of cultural or social imperialism is acceptable', and defended those women who wished to stay at home with their families.[17] As the recession bit, both Giroud and Pelletier promoted the ideas of part-time work for women and schemes for maternity leave, much to the ire of feminists like Simone de Beauvoir, who founded the Ligue des Droits des Femmes to fight them. One concession to women in politics was a bill endorsed by the Barre government in June 1979 which required a quota of 20 per cent of women candidates on lists for municipal elections. But this did not come before the Assembly until November 1980, and was lost as presidential elections loomed.

While many feminists refused to fall into the trap of becoming involved in political parties, others entered party politics to bring feminist demands onto the political agenda. The leaders of the MDF, Audry, Eyquem, and Roudy, helped to found the new Socialist Party in 1971, in the hope that it would be receptive to their ideas. But the Marxism that held sway in the era of the common programme with the PCF put class before gender and regarded feminism as a bourgeois luxury. After the break-up of the united Socialist-Communist front in 1978, however, a Socialist convention accepted a manifesto of women's rights, and on 6 October 1979, anniversary of the march of the women of Paris to Versailles, and as the Veil law came to the end of its trial period, 400,000 women marched in Paris for a permanent abortion law, Socialists and left-wing unions joined by the MLF, Choisir, the Ligue des Droits des Femmes, and Planning Familial.

In the parliamentary elections of 1981, only sixteen of the 269 socialist deputies elected were women. But six women became ministers, and this time of important departments such as agriculture (Édith Cresson), environment (Huguette Bouchardeau), and a junior post at the ministry of defence (Edwige Avice). Yvette Roudy became minister of women's rights—a significant change in title from women's affairs—

and introduced a package of new legislation. The cost of abortion was henceforth to be reimbursed by social security. A law of occupational equality (13 July 1983) outlawed refusal to employ, train, or promote on grounds of sex, and businesses were required to submit annual reports on measures taken to realize equal opportunity. Even women's rights, however, were not exempt from the U-turn of 1983. Roudy's ministry lost its cabinet status, and Mitterrand switched to a natalist policy offering time out of work to either parent in order to raise a third child. A bill promoted by Roudy to allow law-suits to be brought against advertisers who used degrading images of women was hounded out by the advertising lobby, and no action was taken on the report of a commission chaired by Benoîte Groult to officialize the female form of words like deputy (*la députée*) after howls from the Académie Française. A bill sponsored by Gisèle Halimi, now a Socialist deputy, to impose a women's quota of 25 per cent on lists at municipal elections was passed by the National Assembly but overturned by the Conseil Constitutionnel on the grounds that it infringed the Declaration of Rights of Man. A conference organized by Choisir in October 1983 was therefore aptly entitled *Is Feminism Finished?*

Persistent inequalities

Some women became stars by conquering male bastions. In 1972 the entrance examination for Polytechnique was opened to women and Anne Chopinet came top; unfortunately she was not allowed to wear a sword in the march-past on Bastille Day 1973 and carried the flag instead. Danielle Decure became the first woman pilot in 1975, Marguerite Yourcenar the first woman elected to the Académie Française in 1980, and Sylvie Girardet the first woman, stockbroker in 1985, even though in theory the Bourse had been open to women since 1967.

These successes, however, only served to highlight the discrimination suffered by women in professional life. This was not a reflection of their educational attainment. In 1992, 53 per cent of those passing the *baccalauréat* and 55 per cent of those in higher education were women. More women than men passed the *agrégation* in 1982, but only 8 per cent of university staff recruited that year were women. In 1989 only 10 per cent of those admitted to the scientific *grandes écoles* were women, only 2 per cent higher than the figure for 1972, and only 7 per cent of *ingénieurs* were women. In 1992 women were almost 30 per cent of managers, but in one business out of two women managers were less than 10 per cent of the total and overall female managers earned 30 per cent less than their male colleagues. Marriage, which invariably helped men in their careers, acted as a brake on the careers of women. A survey of 1987 showed that to become senior managers women had to have had more education than men: half of female senior managers had done three years of higher education as against a third of their male colleagues. This discrimination served as an incentive to highly educated women to remain single, which was the status of 33 per cent of French women aged 20–40 in 1989. At the same time the movement to encourage women to stay at home and raise families did not seem to have weakened. An allowance available to a parent who wished to leave work to look after the third child was made available by the Chirac government in 1987. Simone Veil, as minister for social affairs, promised to make it available from the second child in 1993, in response to warnings from Colette Codaccioni, RPR deputy for the Nord, that 'the French family is in danger . . . France has no more children, France is dying.'[18]

A second problem, that of the massive under-representation of women in political life, seemed no nearer a solution. Only 33 women were elected to the National Assembly in 1988 and 35 in 1993. This was the lowest score of any European country with the exception of Greece. Édith Cresson was appointed France's first woman prime minister in 1991, but lasted less than a year. As the 50th anniversary of women securing the vote approached, discontent about this record became focused and an organization, Réseau Femmes pour la Parité, was set up. They demonstrated outside the National Assembly after the elections of 1993 to demand 'liberty, equality and parity'. The Réseau also campaigned for the transfer to the Panthéon of the remains of Olympe de Gouges, author of the Declaration of Rights of Women in 1791 and guillotined in 1793. François Mitterrand conceded that he might transfer the (non-political) ashes of Marie Curie, to which the Réseau replied ironically that the pediment would henceforth have to read: 'Aux grands hommes et à une femme, la Patrie reconnaissante'.[19] Meanwhile dissident socialists Gisèle Halimi and Jean-Pierre Chevènement put the idea of parity into action in the Mouvement des Citoyens list for the European elections of 1994, by including equal numbers of women and men. However, they were allowed only 96 seconds of air time to publicize their views, and obtained less than 3 per cent of the vote.

The final obstacle in the way of the equality of women was an ingrained sexism in French society, matched by a refusal of women themselves successful and liberated to take feminism seriously. Most obvious was a vociferous and active anti-abortion movement, which refused to pay taxes that might be used to reimburse abortions, demonstrated outside the laboratories of a firm on the boulevard des Invalides in January 1988 to denounce the 'morning after' pill it had developed, and occupied clinics where abortions were performed until a law of December 1992 criminalized such acts and hit-squads were sent for trial. Another indication of this sexism was the brushing aside of the Groult commission's attempt to legitimize feminine equivalents of conventionally masculine titles or trades. The effect of this was simply to imply that women who did accede to such posts were invisible. The destruction of the anti-sexist bill was an overt manifestation of an attitude so pervasive in French society that it was found even among leading intellectuals. In 1993 Bernard-Henri Lévy, himself married to a film star, replied to Françoise Giroud, concerned that coffee could not be advertised without showing a nude woman in ecstasy, 'Long live nude women in ecstasy!'[20] While that might be expected from a self-adoring male, it was more disconcerting to learn from the radical critic Hélène Cixous, interviewed in 1992: 'If, in France, I do not call myself a feminist . . . I sometimes choose, on the other hand, on other foreign soils, to call myself a feminist.'[21] If a thinker like Cixous could not bring herself to call herself a feminist in France, what hope was there for feminism?

Jacobins and regionalists

It was written into the constitutions of the Fourth and Fifth Republics, as it had been in that of the First, that France was a Republic One and Indivisible. That there should be one legislative body, one centralized administration, and one revolutionary ideology was the orthodoxy of Jacobins, a breed as familiar in the France of 1945 as in that of 1793. To suggest anything else was, for Jacobins, to play into the hands of counter-revolutionaries, who at worst wanted to revive the provinces of the Ancien Régime, with their noble-dominated assemblies, at best wanted to free municipal councils and departmental *conseils généraux* from the grip of prefects, the all-seeing agents of the centralized government. There had been, from time to time, attempts to revive the provinces in France, usually called regions in order to sidestep the accusation of counter-revolution, in order to protect provincial languages, culture, or history from the centralized administration and ideology beloved of the Jacobins. This had particularly been the case in Brittany, Flanders, Corsica, and Alsace and Lorraine after their recovery by France from Germany in 1918. There had also, more successfully, been attempts to develop local democracy, giving more power to municipal councils and *conseils généraux* and less to prefects. Even here, however, Jacobins had been keen to denounce federalism and accuse the decentralizers of insidious counter-revolutionary projects. It was essential for them that prefects should control the localities, and put their influence at the disposal of republican politicians seeking election or re-election.

The Jacobin argument seemed to be strengthened as a result of the Occupation of 1940–4. The defeat of the centralized Jacobin state, the presence of German (and, in Corsica, Italian) forces, and the existence of a state at Vichy that cherished all sorts of reactionary ideas, opened up all sorts of opportunities for regionalists. A minority, particularly in Alsace and Brittany, pushed for nothing less than separatism and hoped that the occupiers would offer them autonomy within the new world order. Unfortunately for them, German strategy was determined by strategic considerations that included firm control of the French coast, the administration of Flanders from Brussels, and the re-annexation of Alsace-Lorraine. Other regionalists looked to Marshal Pétain's régime at Vichy, which heaped scorn on Jacobinism and preached the virtue of France's ancient provinces, in order to gain concessions. Breton regionalists, for example, obtained a Breton Consultative Committee in 1942, which then petitioned for a provincial assembly with financial and legislative powers, a Breton executive and joint status for Brittany as an official language. Vichy, however, was no more prepared to placate the regionalists than previous regimes, and even strengthened administrative centralization by establishing a network of regional prefects, above the departmental prefects, in 1941.

As a result of their flirtation with Vichy, Italians, and Germans, regionalists were totally discredited after the Liberation. A few were executed as collaborators, others were sentenced to prison terms or went into exile. The provisional government re-established the One and Indivisible Republic with a vengeance; even the *commissaires de la République*, sent in by de Gaulle in 1944 to establish his authority in the same areas controlled by Vichy's regional prefects, were abolished in 1946 in order to restore the cosy duopoly of local deputy and departmental prefect. Further, under the Commissariat au Plan the regional policy of the Fourth Republic was to dissolve old-fashioned regionalist sentiments by attacking what was seen as its root cause: the relative backwardness of the peripheral parts of France. Under the Fifth Republic regional policy was renamed 'l'aménagement du territoire' (regional development), a technocratic term that meant nothing more than the rapid modernization of those parts that lagged behind, not to promote regionalism but to ensure that it withered away. In 1964 regional prefects were brought back to supervise these programmes. Each was assisted by a Commission de Développement Économique Régional (Regional Economic Development Commission, or CODER), which gave no voice to regionalists but simply mobilized local politicians and business leaders in a consultative capacity behind the regional prefect.

Regionalism did not, however, go away; it simply changed its spots. It revived from 1960s, with a different ideology, a different means of action, and to some extent in different regions. Alsace-Lorraine, which had been the leading edge of regionalism between the wars when the Jacobin state, having recovered it from Germany, tried to impose the French language and the anticlerical legislation of the pre-1914 period on it, no longer constituted a problem. The most popular party there, the MRP, was also one of the ruling parties of the Fourth Republic, and was able to negotiate a 'special status' in respect of its church schools, state funding of the Church, and teaching of German. Moreover, after 1945 Alsace-Lorraine was no longer a battle-ground between France and Germany but the centre of the European Community, part of the heartland of its prosperity and the home first of the Council of Europe, then of the European Parliament.

The regionalist offensive now came from other parts of France. Brittany had always been a problem, but now the south was the main platform of regionalism. The French Basques, who numbered only 230,000, a tenth the number of Spanish Basques over the border, became infected by the separatist struggle of their Spanish brethren against Franco's state, especially as the French state helped Franco to round up militants and bring them to justice. Occitanie, essentially Languedoc and Provence, was invented as a region with its own identity and claims to autonomy virtually from nothing in the 1960s. More focused and more serious was Corsican regionalism, which was founded on a tradition of independence between Genoan rule and French conquest in 1769, and which became increasingly violent.

These regions were poor in comparison to the rich north and east of France, much closer to the core of the European economy. The south, moreover, was historically the most revolutionary part of France, and the new regionalism of the 1960s evolved a progressive, even revolutionary regionalism that was quite different from the traditional regionalism that had held sway before 1945. It took inspiration from the wars of decolonization, and in particular from the FLN in Algeria and Castro's Cuban revolution of 1959. It argued that the poor south and west of France were 'internal colonies', maintained to serve the booming economies of the rich north and east, and that regionalism had to make common cause with socialism and defend its assets by taking them into regional ownership. In this way regionalism both recovered legitimacy and linked up effectively with parties and trade unions opposed to the Gaullist and Giscardian Republic. The traditionalists, for their part, were also reworking their argument in order to shake off the stigma of reaction. Though nationalism and ideas of racial superiority had been discredited by association with Vichy and Nazism, the idea of the ethnic group oppressed by a dominant nationality was full of possibilities, and was indeed given the blessing of Pope John XXIII in his encyclical *Pacem in terris* (1963). Traditionalists, armed with the concept of the ethnic group, now felt able to refute the Jacobin claim that the French were all, as citizens, members of the same nation and that ethnic differences counted for nothing.

Both traditionalist and progressive arguments were mobilized in those parts of France where regionalism was strong. In Brittany the traditionalists were headed by Yann Fouéré, who had been secretary-general of the Breton Consultative Committee under Vichy and returned from exile to set up a Mouvement pour l'Organisation de la Bretagne (Movement for the Organization of Brittany, or MOB) in 1957. This reiterated the old regionalist argument that the terms of the treaty of 1532 by which the Breton nation had been united with France, and which had been torn up at the Revolution, must be honoured in a new federalist France, according Brittany a regional assembly to manage its own affairs and recognizing Breton as an official language alongside French. The progressive wing was represented by the Union Démocratique Bretonne (Breton Democratic Union, or UDB), founded in 1964. It popularized the slogan 'Brittany = colony', supported the struggles of Breton peasants and workers, defended the Breton language and environment, and worked closely with the CFDT and later the Parti Socialiste.

The same coexistence of traditional and progressive regionalism was found in the south. Robert Lafont, who developed the concept of the internal colony and advocated a combination of regionalism and socialism, founded the Comité Occitan d'Études et d'Action (Occitan Committee for Study and Action) in 1962 and claimed the revolutionary heritage of the Languedoc wine-growers' revolt of 1907, the 'Midi rouge' of 1848 and 1871, the Huguenots, and the Cathars who had been subjugated by the Albigensian crusades of the thirteenth century. For the traditionalists François Fontan, who founded the Parti National Occitan (Occitan National Party) in 1959, argued that the Occitans were an ethnic group that had suffered genocide at the time of the Albigensian crusades. Among the French Basques Enbatu, founded in 1963, was traditionalist, while progressive regionalists founded the Basque Socialist Party in 1975. In Corsica, where a development plan of 1957 threatened to reduce the economy to a wine monoculture and tourism on the Majorcan model, the traditionalist response was orchestrated by Action Régionaliste Corse (Corsican Regionalist Action, or ACR), founded in 1967 by the brothers Edmond and Max Simeoni and demanding internal autonomy and the recognition of the Corsican 'ethnie' or 'people'. Meanwhile, the progressives of the Front Régionaliste Corse (Corsican Regionalist Front), set up in 1966, imbibed the ideas of Robert Lafont and demanded regional ownership in order to end internal colonization. Both demanded a Corsican legislative assembly in order to defend Corsican interests.

The strategy of all these movements was success within the democratic system. The system, however, not only was democratic but relied on the bureaucracy, police, and in the last resort the military to uphold it. Moreover, local politicians in Corsica, the Basque country, and Brittany, as anywhere else in France, found it in their interests to join national parties and co-operate with the administration in order to obtain favours for their constituents and ensure their re-election. Michel Labéguerie, a member of Enbatu, was elected deputy in 1962 in the absence of a Gaullist candidate, but soon joined the Centre Démocrate and went on to become a minister. Given these obstacles, some regionalists felt that direct revolutionary action or terrorism was the only way to achieve their goals. The technique was nevertheless propaganda to alert opinion, rather than the destruction of the state. The Spanish Basque movement ETA, founded in 1959, showed the way, although there was no French Basque terrorist organization until Iparretarak ('Those of the North') in 1981. A Breton terrorist organization, the Front de Libération de la Bretagne (Breton Liberation Front, or FLB), was formed in 1966 to demand total independence and attacked symbols of French oppression such as tax offices, or of exploitation, such as the villa of the Paris property speculator Bouygues. The Front de Libération Nationale de Corse (Corsican National Liberation Front, or FLNC), set up in 1976, likewise demanded complete independence rather than the internal autonomy desired by the ARC, and the number of bomb attacks rose from 43 in 1973 to 462 in 1980 and 742 in 1982.

Towards a Decentralized Republic?

'How can you govern a country that has 246 varieties of cheese?' once asked General de Gaulle.[22] His mission had always been to fortify the French state, and he had no intention of undermining the One and Indivisible Republic. The regional prefects of 1964 represented a return to the intendants of the Ancien Régime, and the CODER served the interests of local politicians and business leaders eager for planning concessions, not regionalist campaigners. The regional assemblies offered to—and rejected by—the French people in the referendum of April 1969 were no better than the CODER: they added deputies and senators to the local politicians and business leaders, were confined to economic, social, and cultural matters, met only twice a year, had no permanent staff and next to no budget, and were under the firm control of the regional prefect. The regional councils set up by Georges Pompidou in 1972 were even more of a gift to the political class. They were reserved for local and national politicians, while business leaders were hived off to an economic and social committee.

While marginalizing regionalist leaders, the Gaullist and Giscardian regimes dealt severely with regionalist militants. The FLB was dismantled in 1968–9 on the eve of an official visit of de Gaulle to Brittany. Reconstituted, it resumed its attacks, and eleven FLB militants were tried by the Cour de Sûreté de l'État in 1972, eight of whom were sentenced to prison terms. In August 1975 French troops occupied Corsica in an attempt to put an end to regionalist demonstrations, the ARC was dissolved, and four Corsican militants were sent to prison early in 1981 by the Cour de Sûreté de l'État.

The Socialist opposition, at first blush, was prepared to relax the Jacobin grip and make significant concessions to regionalists. During the presidential election campaign of 1981, François Mitterrand visited Corsica and promised it 'special status'. Gaston Defferre, minister of the interior and decentralization, sponsored a law that gave Corsica a regional assembly and its own executive under the president of the assembly. In two important respects, however, concessions fell short of Corsican demands. First, the assembly was not legislative but consultative, as there was to be no departure from the orthodoxy of a single legislature in Paris. Second, the Corsicans were not recognized as a separate ethnic group, only as the 'Corsican people, a constituent part of the French people', and even this term was later ruled out by the Conseil Constitutionnel, for the unity of the French nation could not be brought into question. Meanwhile the Corsican assembly in no sense turned out to be a plaything of the regionalists. In the elections to the assembly in August 1982 the Union du Peuple Corse (Union of the Corsican People, or UPC), which had replaced the ARC, won only 10 per cent of the vote and seven seats, and had to join forces with the Radicals in order to elect the president. Subsequently, the government effectively ignored the assembly, provoking the FLNC to renew its campaign of violence and the UPC to withdraw from the assembly.

The Corsican legislation was a pilot test for the law of 2 March 1982, which was the most important measure of administrative decentralization since the Revolution. The great innovation was the introduction of direct universal suffrage for regional councils, to be held every six years from 1986. Power was shifted from prefects and regional prefects to presidents of *conseils généraux* and regional councils, who were now elected, together with their own executives, from the councils. An attempt was made to limit the pluralism of the political class by making deputies, senators, and mayors of towns of over 20,000 inhabitants ineligible for regional councils. Having said that, it was the established parties, and not regionalist movements, that benefited from elections to the regional councils. Regionalists were also totally eclipsed by the National Front, which obtained 10 per cent of the vote in 1986 and 14 per cent in 1992. Regionalists made an impression only where they allied with other parties, most notably the Green Party. Thus in the municipal elections of 1989 an alliance of regionalists and Greens won 15 per cent of the vote at Lorient in the Morbihan, while in the European elections of that year a regionalist–Green coalition won 15 per cent of the vote in Corsica and Max Simeoni was elected to the European Parliament at Strasburg, a seat that he lost in the European elections of 1994. The extent of the impact of regionalists generally may be gauged by their performance in those elections. In Brittany they polled just over 1 per cent, in the Basque country 1.6 per cent, but nationally their score was a mere 0.4 per cent of the popular vote.

After the fall of the Socialist party from power in 1993, the whole issue of decentralization disappeared from the agenda. Under Minister of the Interior Charles Pasqua the regional question was once again recast in the technocratic and bureaucratic terms of 'l'aménagement du territoire'. The referendum on Maastricht had thrown up a split between prosperous and less prosperous cities and regions: the former in favour of Europe, the latter against. Pasqua drove a bill though the National Assembly in July 1994, both to redistribute resources between rich and poor communities and to pick up the pieces remaining after a decade of decentralization, re-establishing the state as arbiter between towns, cities, departments, and regions, without returning to the worst of administrative centralization. This law almost wrecked the right-wing majority, such were the reservations of the political class that had become used to greater self-government. At the same time it totally excluded the idea of regional identity and autonomy, and therefore did nothing to alleviate the festering problem of Corsican nationalism, which continued its violent struggle against the French nation-state.

Believers and immigrants

France considered herself a nation-state and the government of the nation-state was not only One and Indivisible but lay. That is to say that the state was secular and neutral between religions, none of which was officially recognized or funded. Religious marriages were cel-ebrated, but the state recognized only civil marriages performed by civil officials. Churches, synagogues, and mosques flourished, but there was no religious instruction in state schools and individuals were not allowed overtly to proclaim their religious affiliation there. While this was upheld on the grounds of strict equality, members of religious communities, including Catholics, often felt that they were treated as second-class citizens because the needs of their religion were not considered by the state. This was complicated by the question of immigration, for while Catholics were overwhelmingly French by origin, Jews and above all Muslims were not. That said, the problem posed by immigration to the nation-state was far greater than a religious issue. It triggered off controversy about how far immigrants could or should be assimilated into the nation-state, about the relative superiority or inferiority of races, and about the nature of French national identity itself.

Catholics

In terms of religious practice, Roman Catholicism was clearly on the decline in the post-war era. The percentage of children baptized in France fell from 90 in 1945 to 63 in 1975 and was predicted to be 50 in the year 2000. The proportion of Catholic marriages remained above 75 per cent until 1972, then fell to 55 per cent in 1987. While 32 per cent of French people attended weekly mass at the Liberation, only 10 per cent did so after 1980. Recruitment to the Catholic clergy also suffered after the war, so that for every 10,000 inhabitants there were fourteen priests at the beginning of the century and seven in 1965; at the beginning of the next century there would be one.

Paradoxically, however, 81 per cent of French people called themselves Catholic in 1981, as against 15 per cent with no religion and 3 per cent belonging to other religions. To the vast majority of these people Catholicism meant not church-going or even participation in Catholic rites of passage, but belonging to a community with a common culture and a common system of beliefs, constructed over the course of French history. After a century or more of anticlerical persecution, moreover, Catholics were fully integrated into the political mainstream, and exercised considerable leverage over political decisions relating to education.

Catholic schools educated 23 per cent of secondary school pupils steadily between 1958 and 1980. The MRP had failed to secure *la liberté de l'enseignement* or equal rights for Catholic education inscribed in the constitution of 1946, but state funding of Catholic schools, terminated by the Third Republic in 1886 and briefly reintroduced by the Vichy regime, was put on a firm footing by the Loi Debré of 1959. The strength of the Catholic lobby was demonstrated in 1984, when anticlerical socialist deputies amended a government school bill effectively to confine state funding to state schools, and on 24 June a million Catholics, with the full support of the hierarchy and the participation of right-wing politicians like Jacques Chirac, demonstrated in Paris against the bill. President Mitterrand promptly withdrew the bill, and provoked the resignation of the education minister Alain Savary and the prime minister Pierre Mauroy. A limit to the privileges Catholics could extract from the republican state, even under a right-wing government, however, was revealed ten years later. In opposition to a law passed by the Balladur government to take the lid off the amount local authorities could fund Catholic schools, fixed by the Loi Falloux of 1850, up to 900,000 people demonstrated in the rain and (with the help of the Conseil Constitutionnel, which ruled against the law) safeguarded the primacy of lay state education, the 'republican school'.

Jews

Jews were granted civil and political rights in France at the time of the French Revolution. As a rule, they internalized the message that to become model citizens they had to assimilate by putting their Judaism on one side, or at least to keep it to a very private sphere. They espoused the French Enlightenment and fertilized the French economic, cultural, and political élites. Calling themselves 'Israelites', they looked down on the poor, uneducated Jews who arrived in France in the late nineteenth and early twentieth centuries, fleeing persecution in Tsarist, then Bolshevik Russia, the successor states of Eastern Europe, and Nazi Germany. But France's claim to be the land of the rights of man and the most generous of fatherlands was betrayed by the Vichy regime, which purged the Jews from public life and then sent 75,000 of them, both foreign and French, over a quarter of the Jewish population in France, to their deaths at Auschwitz.

The reaction of the survivors at the Liberation was not to protest but to assimilate even more, seeking French nationality if they did not have it, marrying non-Jews, even changing their names. But in time other factors prompted a revival of the Jewish religion and culture. The independence of French North Africa brought an influx of 145,000 Jews from Algiers and Oran, Tunis and Casablanca, bringing the total in France to 360,000 in the 1960s, 535,000 in the 1970s. Far more than Jews born in France, or even in Central and Eastern Europe, these were practising Jews, who observed Jewish customs and read Hebrew. The Six-Day War in 1967, and de Gaulle's anti-Jewish comments at the time, brought home the threat to the state of Israel. The young generation of Jews, having participated in the events of 1968, then reacted against what they saw as the shameful assimilation of their parents, and, under the guidance of the philosopher Emmanuel Lévinas, learned Hebrew to promote a revival of Jewish religion, enlightenment, and culture. This revival may have affected only 20 per cent of Jews in France, but it was an articulate and dynamic minority. Unfortunately, it promoted an equal and opposite reaction of anti-Semitism, expressed in such atrocities as the bombing of a synagogue in the rue Copernic, Paris, in October 1980, the gun attack on a Jewish restaurant in the Marais in August 1982, and the desecration of Jewish graves at Carpentras in May 1990 and at Perpignan in June 1993. The message of anti-Semites, 50 years after the Holocaust, was that any manifestation of Jewish specificity in France was unacceptable.

Muslims

The Arab population of France was an immigrant population from formerly French North Africa. The first wave had come over as single men, to help in the post-war reconstruction of France. After the frontiers were closed in July 1974 the only immigrants allowed in from North Africa (and setting aside illegal immigrants) were women and children permitted under the policy of family reunion. From this moment fewer North Africans returned home, and the immigrant population became sedentary and resident. By 1981 over 70 per cent of foreigners living in France had done so for over ten years.

How effectively these populations would integrate into French society was of the greatest importance. Many of them were or became French nationals. Those who had been born in the Algerian departments before 1 January 1963 were and remained French. Under a law of 1889, passed when France desperately needed fighting men, children born in France of foreign parents automatically became French nationals at their majority, unless they specifically declined to do so. The rate of mixed marriages increased, although they accounted for under 15 per cent of marriages by Italians, Spaniards, Portuguese, and Algerians. The housing of immigrants improved, as they moved from the late 1960s out of shanty towns and hostels for single workers built hastily after 1956 by the Société Nationale de Construction de Logements pour les Travailleurs Algériens (National Company for the Construction of Accommodation for Algerian Workers, or SONACOTRA) into HLMs and high-rise estates. Unfortunately these estates, in the outer suburbs of the great cities, like Les Minguettes outside Lyon, were progressively abandoned by the French residents and became decayed ghettos for unemployed immigrants. In the early years, when two-thirds of North Africans began in France as unskilled workers, the trade unions and Communist party served as effective levers of integration into French society. Of their children, taking the cohort born before 1968, however, only a third remained in unskilled work, and 54 per cent were in clerical or management posts. Immigrants were often accused of bringing down standards in education, but in Créteil outside Paris in 1989, a rural college with 0.25 per cent of foreign children had a pass rate of 44 per cent for the *brevet*, while an urban college with 70 per cent of foreign children achieved 48 per cent.

If one force acting on immigrant communities was the pressure to integrate with French society, another was the desire to integrate their own communities, dispersed in an alien environment. This helps to explain the low proportion of mixed marriages. It also explains the importance given by North Africans and Turks to the establishment of the Muslim religion on French soil. Islam provided a compensation for immigrants who found assimilation difficult or objectionable, and wished to assert the dignity and specificity of their community in the eyes of God if not in those of the French. It was vital to Muslims to assert the transcendence of their religion, with its fasts, festivals, and times for prayer, but to assert it against the constraints of the factory and living space was a struggle. North Africans organized a rent strike in SONACOTRA hostels in 1975 in order to obtain prayer rooms. They petitioned at Renault-Billancourt in 1976 and went on strike at

Citroën-Aulnay-sous-Bois in 1982 to secure time for prayer during working shifts. From 1981 Muslim fathers set up prayer-rooms in HLMs which would also serve as Koranic schools for the children in the evenings and during school holidays. After the law of October 1981 allowed foreigners to form associations, private places of worship became public, and with the help of oil money channelled through the World Islamic League mosques started to be built. Not all had minarets, but whereas in 1970 there had been only 11 Muslim places of worship, by the end of the 1980s there were nearly 1,000, including 73 sizeable mosques and 5 cathedral-mosques, 3 in Paris, 1 in Marseille, and 1 in Lille. Muslim leaders, such as Sheik Abbas of the Paris mosque, argued that the practice of Islam with or without mosques and loyalty to the secular French state were in no sense incompatible. One of his lay assistants, Professor Arkoun of the Institut d'Études Islamiques and the University of Paris III, whose grandfather had fought for the French in 1870, affirmed that 'our aspiration is to be totally integrated while preserving our Islamicness, without which our Frenchness would be imperfect.'[23]

Not all Muslims shared this optimism. A particular problem was the generational conflict between parents and children. The second generation, nicknamed Beurs ('Arabe' in Verlan, a Parisian slang which inverts syllables), acquired French nationality at the age of 18 unless they specifically declined it. Schooled with children of French stock, going out with them, speaking French rather than Kabyle, wearing T-shirts and jeans, exposed to the same mass media as their peers, they found it difficult to be both Muslim and French and tended to abandon religious practices and (in the case of girls) the veil. As one Algerian father, a hairdresser, put it, 'We have lost our children.'[24] The Beurs were keen to assimilate, but trapped on forbidding estates, facing a high rate of unemployment, generally discriminated against, they were clearly not the equals of their French peers. In the autumn of 1983 a Beur march took place from Marseille to Paris, led by Toumi Djaïja, who had been wounded in confrontation with the police in Les Minguettes, and saw himself as a new Gandhi or Martin Luther King. Using the slogan 'For Equal Rights, Against Racism', they managed to extract from the government a new ten-year residence permit for foreigners, automatically renewable.

The cause of the Beurs was taken up with a blast of publicity by SOS-Racisme in 1984. Its organizer was Julien Dray, a Jew whose parents had come to France from Oran, a former Trotskyist and now member of the PS, while the front man was Harlem Désir, born in

Guiana of a Martiniquais father, but educated in France from the age
of 14. Designed to build on anti-racism in schools and universities, it
was advised by the clown Coluche and the intellectual Bernard-Henri
Lévy, used all possible devices of the media from a logo of a yellow
hand inscribed 'Touche pas à mon pote' (hands off my buddy) to a rock
concert on the place de la Concorde on 15 June 1985 and an interview
with Harlem Désir on television's *L'Heure de vérité* in August 1987. The
movement celebrated a multireligious, multicultural, multicoloured
French society, and seduced Mitterrand into relaunching the idea of
votes for immigrants in local elections. But the Beurs soon grasped that
SOS-Racisme was in fact a front organization of the Socialist party,
designed only to corral votes, and withdrew their support.

'Assimilation for young Algerians was not an easy choice and did not
become any easier. It threatened to deprive them of their identity and
dignity, while leaving them eternally as second-class citizens. For
assimilation was a condition of future rights, not a guarantee of them.
The numbers of immigrants requesting French nationality declined
after the immediate postwar years. Algerians were proud of the nationality they had conquered in 1962, often held dual nationality, and
increasingly in the 1980s chose to do their military service in the
Algerian army rather than in the French. It was generally felt that
French nationality would gain them little in view of the colour of their
skins, except possibly security from expulsion, and that after 1984 the
renewable ten-year residence permit was enough. If they were foreigners, they were not allowed to vote, even in local elections. If they were
French nationals they rarely bothered to register, until a registration
campaign spearheaded by an organization called France Plus for the
municipal elections of March 1989, which saw 500 Beurs and 'Beuret-
tes' elected as municipal councillors.

Even Algerians who were not French nationals were subjected to the
pressure to assimilate. The republican school, which had excluded
Catholic religious practice from one door, was not going to allow Islam
to enter by another. Muslims, on the other hand, though keen to have
their children educated, wanted to defend their religious beliefs and
identity. For them, religious values and the honour of the family were
carried by the women, but in October 1989 three teenage Muslim girls
at a college in Creil (Oise) were refused admission to class for refusing
to removed their Islamic headscarves. The college principal cited the
obligation of the republican school to remain entirely secular and
prohibit all forms of proselytism within it, although wearing discreet
emblems like a crucifix or star of David was acceptable. The rector of

the Paris mosque demanded that all religions be treated equally, the ambassador of the Arab League in France denounced the persecution of Islam, and rabbis and bishops joined in a demand for tolerance of signs of religious faith. The attitude of the authorities was incoherent, on the one hand determined to impose the religious neutrality of the republican school, on the other seeking to avoid undue controversy. On this occasion the Socialist minister of education, Lionel Jospin, agreed that the girls should return to school. Five years later, however, in September 1994, a circular of the Centrist education minister, François Bayrou, reiterated that 'ostentatious signs' of religious allegiance were prohibited from schools, and on the basis of this the head of the Lycée Faidherbe in Lille banned twenty Muslim girls wearing headscarves.

For many Muslims it was now made clear that the price of assimilation was too high: their religion and identity were at risk. Thus the period after 1989 saw a revival and radicalization of Islamic belief both in poor Algerian suburbs and among educated and articulate Algerian youth, who at one time had looked like abandoning the faith. Organizations were set up by the Muslims to fight drugs and crime, to teach the faith, and to defend the wearing of veils. Sympathy with the fundamentalist Islamic movement in Algeria was expressed by Algerian Fraternity in France which, at the time of the killings of five Frenchmen in Algiers in August 1994, was denounced as an offshoot of the Islamic Salvation Army and had its leaders arrested. The confrontation between disillusioned Algerian youths, returning to fundamentalism, and the police state of Charles Pasqua, rooting out terrorism, made reconciliation between the Muslim community and the French state less likely than ever.

National identity and racism

The issue of Islamic scarves highlighted a debate about French national identity and strategies for dealing with immigrants of different religious and cultural backgrounds. The French, unlike the Germans, did not consider themselves a *Volk*, bound by ties of blood, soil, and language. Whereas the Germans constructed themselves as a nation before they became a united state, France was forged as a centralized state by successive kings, while at the Revolution the French people made a new social contract to enthrone themselves as sovereign, the source of all law, with all French citizens equal under the law. The nation was thus the body of French citizens who made and obeyed the law, and had been constituted as such by an act of will. This rhetoric, that nationality was acquired by an act of will, clashed, however, with nationality legislation as formulated after 1889. This specified that two basic criteria for French nationality were blood (parentage) and soil (residence). Moreover, the French considered that in order to become a French citizen an individual had to assimilate French civilization, in particular its language. The task of the republican school was to impose this civilization. Finally, it was argued that assimilation was possible only for individuals, not for communities, and that while it was permissible to practise religion and ethnic traditions in private, there could be no public recognition of ethnic or religious communities which might assert claims against and fragment the French nation-state.

There was no recognition of the *droit à la différence* in France until teaching of the mother-tongue of immigrant communities in school, at the expense of the country of origin, was allowed under bilateral agreements with Italy and Spain in 1973 and with Algeria in 1981. This represented the emergence of a more liberal strand of thought which accepted pluralism in civil society and thus the possibility of multicul-turalism. The term 'assimilation' tended to give way to 'integration'. The immigrant populations, however, were not keen to take up these classes, which acquired the reputation of little ghettos. Moreover, such discrimination played into the hands of racists who argued less that races were unequal than that national and racial differences were objective and insurmountable and that assimilation would never work, in order to justify their argument that immigrants should be sent home.

The argument that immigrant groups could not be assimilated was made with particular force in the case of North Africans. They differed from French stock in two fundamental ways: race and religion. They were 'Arabs' and Muslims and as such threatened to defile the French nation and undermine its Christian civilization. 'Islam', said Jacques Soustelle, the veteran defender of French Algeria in 1990, 'is not only a religion, a metaphysics and ethics, but a determining and constrictive framework of all aspects of life. Consequently, to speak of integration, that is to say assimilation, is dangerously utopian. You can only assimilate what can be assimilated.'[25] The revival of Islamic fundamen-talism after the Iranian revolution of 1979, and the appearance of the Islamic Salvation Front in Algeria in 1989, served to reinforce such views. These attitudes were also nurtured by painful memories of the Algerian war. It should not be forgotten that many militants in the Front National, from Le Pen downwards, had fought in the Algerian war or were *pieds-noirs*. For them, during the period of colonialism,

Algerians had not been considered equal citizens unless they renounced Muslim law or until they rose in revolt. They were seen as terrorists and traitors, and reminders of the failure of France's civilizing mission, indeed of its Empire. The resurgence of Islamic terrorism in Algeria in 1993–4, when 57 foreigners were assassinated within a year, and the hijacking of an Air France Airbus in Algiers over Christmas 1994, when three hostages were killed, only served to confirm the prejudice that Algerians were terrorists at heart.

While the French had left Algeria, however, they had brought the colonial problem home with them in the form of immigration. Now the colonists were themselves being colonized. Unable to accommodate immigrant communities, and in particular the Algerians, the French were quick to blame them for the ills of society in general. They were held responsible for unemployment, overcrowding, dirt, crime, AIDS, and social-security scrounging. If the native French had been able to look down on the immigrant populations, their racism might have been less pronounced. But as the recession threatened them with social *déclassement* so they saw immigrants doing well in school, taking their jobs, dressing smartly, and driving BMWs. 'Quite honestly,' said Le Pen in 1990, 'I have the feeling that I am being fundamentally persecuted myself.'[26] Since they were losing social superiority over the immigrants they played their last card, which was to assert racial superiority. And they argued that immigrants could not be assimilated at precisely the moment when there was a chance that they would be assimilated.

The politics of race

The politics of race revolved around two issues, immigration and naturalization. As a rule, the Left tended towards a more liberal policy and the Right was more hard-line, not least because after 1981 it was competing for votes with the Front National. The constituency that thought that the ideas of Jean-Marie Le Pen were right, even if it would not vote for him, was very wide. This had an impact on Communists in the red belt around Paris and in other large cities, who feared losing votes to the Front National. It also frightened the Socialists, who increasingly adopted hard-line policies, while Jacobins amongst them had always favoured assimilation rather than the toleration of differences.

Once the door had been shut on new immigration, except for family reunions, Lionel Stoléru, secretary of state in the ministry of labour in the Chirac government, launched a scheme to pay a million of the 3.4 million foreigners in France to return to their country of origin. But the 'Stoléru million' turned out to be no more than 100,000, half of them Italians and Spaniards and only a quarter North African. A Franco-Algerian agreement of 1980 secured the return of 50,000 Algerians down to 1983, but in 1981 the Socialist government of Pierre Mauroy promptly issued an amnesty for 140,000 illegal immigrants. This generosity did not last long. Defeat in the municipal elections of March 1983 provoked a U-turn in immigration policy as in every other sphere. Not only were 50,000 North Africans helped to return in the period 1984–6, but the numbers of immigrants expelled rose from 2,861 in 1982 to 12,364 in 1986. Moreover, while Mitterrand had promised as a candidate for the presidency to give the vote to immigrants in local elections, he put this on ice after a poll in 1984 revealed 75 per cent of the electorate against it.

The return of the Right to power in 1986 brought Charles Pasqua, a Corsican with all the finesse of a New York cop, to the ministry of the interior. He introduced a law on the entry and stay of foreigners (9 September 1986) which refused entry without justification of means of existence and made expulsions easier. Television viewers were treated to the sight of 101 Malians being dragged onto a charter plane at Orly. He also drafted a bill to refuse automatic naturalization to children born in France to foreign parents. But this was opposed by centrists as well as Socialists, by President Mitterrand, and by 100,000 students who demonstrated in December 1986, and the bill never reached the Assembly.

The whole question of French nationality went before a commission chaired by Marceau Long, vice-president of the Conseil d'État, in 1987. It listened to evidence from a wide range of pressure groups and community leaders. The president of the right-wing Club de l'Horloge warned that France was in danger of going down the path of multinational decadence like Austria-Hungary, the Ottoman Empire, or India. By a curious inversion, liberals, in order to defend the status quo, argued that blood (parentage) and soil (residence) were the criteria of French nationality. Against them, conservatives argued that if the children of immigrants became French automatically they would be French 'in spite of themselves', or French men and women only on paper. Thus they endorsed the view of the French Revolution that nationality was not given but a matter of will and choice.

The Long commission, which reported in 1988, indeed recommended that children born in France to immigrants should not become French automatically at the age of 18 but should have to request French nationality. The Socialists, restored to power, did nothing to bring forward legislation, and a frustrated Pasqua and his friends tried to ram a bill through the Senate in the dead of night on 20–1 June 1990 to give effect to the report. By then opinion in France about nationalization and assimilation had been substantially altered by the headscarf affair. In the name of both feminism and national identity, many on the Left refused to make concessions to the Muslim girls. Gisèle Halimi of Choisir broke with SOS-racism, which was supporting the Muslims, on the ground that the veil perpetuated the enslavement of women. Lionel Jospin, the education minister, while allowing the girls back to school, said 'I see no reason to change the French model. I am not in favour of substituting the Anglo-Saxon model of communities for the individual French model.'[27] By this he meant that religion should be a private affair and not hinder the assimilation of individuals into the French Republic and nation; there could be no question of integrating communities of which religion or ethnicity was a defining characteristic. Meanwhile a poll of October 1991 showed that 32 per cent of French people agreed with Le Pen's ideas, as against 18 per cent in September 1990.

In March 1993 the Right returned to power with a vengeance. Pasqua, back at the ministry of the interior, wasted no time. First, the Long proposals on nationality were taken up and indeed reinforced, and voted into law in May 1993 despite Socialist accusations of 'apartheid' and 'ethnic cleansing'.[28] A demonstration against it in Paris massed only 1,500 protesters, compared with 100,000 in 1986. The immigrant quarter of the Goutte d'Or in north-east Paris was subjected to a reign of police terror and a law passed, despite objections by centrist ministers Pierre Méhaignerie and Simone Veil, to permit random identity checks of anyone suspected of being a foreigner. Lastly, a new bill on the entry and stay of foreigners was tabled, to make entry more difficult, not least for those seeking asylum, conditions of existence less hospitable, and expulsion easier. SOS-Racism, trade unions, and the Left brought 15,000 demonstrators onto the streets, but the most effective opposition was the Conseil Constitutionnel, which nullified the provisions of the law relating to the right of asylum. An enraged Pasqua demanded that the constitution be revised to plug the loophole, and François Mitterrand caved in. On 19 November 1993, France witnessed the sorry sight of deputies and senators gathering at Versailles to revise the constitution, limiting the rights of man in order to placate racist opinion. Despite weak protests from Socialists, organized labour, and anti-racists, a consensus seemed to be established in France which was content to take a hard line on immigration, the acquisition of French nationality, and the assimilation of the French language and civilization.

6

Cultural Revolutions

The French have always prided themselves as a nation on their intelligence. Their education system, as we have seen, makes a virtue of intellectual élitism. Professions such as teaching, lecturing, the liberal professions, advertising, journalism, and careers in the arts are known as 'intellectual professions'. Certain writers, artists, or academics with a wider role of criticism and moral leadership in society, known as 'intellectuals' since the Dreyfus Affair, have been the objects of public reverence and national pride. But France has also experienced the development of a mass culture in the fields of cinema, television, reading matter, and music. Often American in origin, it has threatened to destroy art and literature and French art and literature in particular, replacing it by a homogeneous culture accessible to all, irrespective of differences in levels of education. In response to this the French state has developed a cultural policy in order to confront mass culture, intended to democratize high culture and defend French culture. These three elements—the rise and fall of the French intellectual, the challenge of mass culture, and the response of official policy—and the tensions between them form the substance of this chapter.

Intellectuals in France

Intellectuals in France were defined against both men of letters and politicians. Against men of letters in their ivory towers, pursuing art for art's sake, intellectuals assumed a public role, questioning and criticizing political and social conditions. Unlike politicians, however, intellectuals drew their authority in the public domain not from having been elected to office but from excellence in the realm of art, literature, or science. Intellectuals were seen to be moral arbiters, not power-seekers; in pursuit of truth, not votes; at the service of universal values such as liberty and justice, not at that of interest groups or political parties. Their contact with the public was secured not through political parties

but through manifestos they collectively signed, reviews they collectively edited, and demonstrations they headed, arm in arm.

   Though they had these features in common, post-war intellectuals did
not form a bloc. Each generation differed from the one before, at once
in its sociological make-up, its cultural concerns, and its relation to the
political issues of the day. The first generation, that of the Liberation,
was dominated by Jean-Paul Sartre, Existentialism, and the vexed
question of relations with the Communist party. After 1956 the
dominant intellectuals were either Marxists who had broken with the
Communist party or structuralists who were more interested in science
than in politics. From 1974 the so-called 'new philosophers' captured
the limelight, but there was also a much wider debate about the role of
intellectuals in modern society and, indeed, whether they had not
suffered a demise altogether.

Sartre and his Generation

At the Liberation there was little doubt as to the identity or doctrine of
the intellectual community in France: it inhabited the Left Bank of
Paris, was led by Jean-Paul Sartre, and preached Existentialism. Some
had posts in the university of Paris or the major lycées; others, like
Sartre, resigned their teaching posts to become freelance intellectuals,
living by the pen. Sartre himself had offices at Gallimard, which
published *Les Temps modernes*, a review he launched with Simone de
Beauvoir and the philosopher Maurice Merleau-Ponty in October 1945,
and which was later joined by Camus. His circle met and worked either
there or in the nearby cafés of the Saint-Germain-des-Prés district, Les
Deux Magots and the Café de Flore, which had the advantage of being
heated at a time when coal was in short supply. They also frequented
cellar-clubs such as the Mephisto and Tabou, where Boris Vian played
the trumpet and Juliette Greco began her career as a singer, Sartre
writing 'La Rue des blancs-manteaux' for her. While Merleau-Ponty
always remained a philosopher, Sartre embodied the intellectual as
polymath. He was a philosopher of substance himself, but was also a
journalist, literary critic, novelist, and dramatist when the vogue was
for philosophical rather than psychological novels and plays and
exposure on the Paris stage conferred an enormous reputation.

   Sartre and his circle insisted on the importance of *engagement* or
commitment on issues of public importance, to demand radical change.
This they did to differentiate themselves from writers like Flaubert and
the Goncourts, whom Sartre held responsible for the Paris Commune
because they had written nothing against it, and from those who had

defended a bourgeois and conservative order that had now been swept away. In October 1945 Sartre gave a lecture entitled 'Existentialism Is a Humanism' to a packed and excited hall in Paris. He attacked the bad faith of the bourgeoisie who had opportunistically accepted the privileges conferred on them by the Vichy regime while marginalizing and excluding others. He also sought to find a bearing in the 'cyclone' created by the Occupation in which all moral and political certainties had been destroyed. In this moral chaos, he argued, it was impossible to cling to so-called eternal values, either Catholic or Kantian. Each individual had the freedom to choose, but also the responsibility to make the right choice. Further, since there was no such thing as human nature or social determinism, each individual had to create and define himself by the decisions he or she made, and people were nothing but the sum total of their deeds. This philosophy was not necessarily that of those who had fought in the Resistance, but was a cult of action that condemned wartime passivity and justified those who had taken the right decision under the Occupation against those who had not. Moreover, it was popular among a generation that was keen to denounce the failings of their fathers and the boy-scout moralism of Vichy and to legitimate a bohemian life style, heavy discussion, late nights, and free love, and was certainly constructed by the press as a doctrine of a nihilistic and amoral youth.

As a philosophy of moral and political commitment, Existentialism came up against the challenge of Communism. Communism, after the brief confusion of 1939, saw itself as the force, embodied in the proletariat and the Soviet Union, that alone had effectively withstood and defeated Nazism, knew the way to the future socialist society, and had the means to realize it. To join the Communist party, the party both of martyrs and victims, must be the right decision, especially for intellectuals who were tainted by their social origin in the bourgeoisie. Existentialism and Communism, however, had an ambivalent relationship. On the one hand Sartre was worried by the latter's historical determinism, its denial of subjectivity and individual choice, and its subordination of culture to the needs of realizing the future socialist society. On the other, Communists were wary of allowing (usually bourgeois) intellectuals an independent role within the party and dismissed Existentialism as bourgeois egoism, oblivious of the forces ruling history or the demands of building socialism. In 1952, however, four years after the failure of an attempt to launch his own political movement, the Rassemblement Démocratique Révolutionnaire (Revolutionary Democratic Union), Sartre moved closer to the PCF, not as a

member, but as a fellow-traveller. The attractions of Communism were
certainty about the future and access to the proletariat, whose liberation was the agenda of history. Sartre took part in the Communist-led
peace movement, visited the Soviet Union with Simone de Beauvoir in
1954, and preached the maxim that 'Il ne faut pas désespérer Billancourt', that all positions should be judged in the light of the interests of
the traditional industrial proletariat.[1] Sartre's love affair with Communism provoked angry divorces from Camus and Aron, and for Sartre
himself the affair soon turned sour. While French Communists accepted
a good deal of Soviet repression in eastern Europe as the price of the
triumph of socialism against imperialism, the invasion of Hungary in
1956 lost the Soviet Union the legitimacy it had boasted since Stalingrad and was opposed both by Sartre and by certain card-carrying
Communists who either left the party or were expelled.

From Marxism to *gauchisme*

Disillusionment with the PCF and the Soviet Union did not necessarily
undermine Marxism, which was too protean an ideology ever to be set
into a single mould. Marxism remained the main ideological basis of
the criticism of contemporary society, developed by intellectuals who
had left or been expelled from the Communist party. Taken up by the
student movement, mixed with other strains of thought in ever more
heady cocktails, it fuelled the protest movement of the 1960s and in
particular the *gauchisme* of 1968.

A first inspiration of the movement of 1968 was the review *Socialisme
ou barbarie*. Launched in 1949 by two Trotskyists, the former Greek
Communist Cornelius Castoriadis and Claude Lefort, a pupil and
friend of Merleau-Ponty, it attacked both Stalinism for bureaucratizing
and betraying the Russian Revolution and the bureaucratization of
advanced capitalist societies. It preached the virtue of proletarian
democracy in workers' councils, from the Russian soviets of 1905 and
1917 to the workers' councils set up in Poland and Hungary in 1956.
Though the review closed in 1965, it was studied by Cohn-Bendit and
the 22 March movement at Nanterre and was one source of their
thinking on workers' control or *autogestion*.

A second inspiration originated in the thought of Henri Lefebvre,
professor of sociology at the university of Nanterre. He had joined the
PCF in 1928, became its leading Marxist philosopher after 1939, and
wrote a stinging attack on Existentialism in 1946. Yet he was marked
by his discovery of the early, humanistic, writings of Marx, which spoke
of the total alienation of man in society, not only of his economic

alienation as a result of wage-labour. This led Lefebvre to a global criticism of everyday life in advanced capitalist society, which he characterized as a 'bureaucratic society of controlled consumption', and the conclusion that the coming revolution would have to transform not only economic relations but social and sexual ones as well.[2] For this heresy he was expelled from the party in 1958, but he had enormous influence on the so-called Situationist International, set up in 1957 by one of his pupils, Guy Debord, a film-maker whose *Société du spectacle* of 1967 was a major attack on the consumer society. Situationism was taken up by the student movement to justify criticism of all forms of oppression, spreading from Strasburg in 1966 to Nantes in 1967 and back to Nanterre in 1968, where it inspired the movement of 22 March. It reached its apogee during the Sorbonne occupation in May 1968, when walls were sprayed with such graffiti as 'I take my desires for reality because I believe in the reality of my desires.'[3]

A third inspiration was a mixture of Marxism and anarchism developed by the Union des Groupes Anarchistes Communistes and their review *Noir et rouge*, which was launched in 1956. Sheltering under the umbrella of the main anarchist movement, the Fédération Anarchiste, they rediscovered the writings of the anarchist Bakunin through the anthologies and commentaries on anarchism by the libertarian socialist Daniel Guérin. The synthesis of anarchism and Marxism they developed had them expelled by the anti-Marxist purists of the Fédération Anarchiste, but their numbers swelled in 1967 and 1968 as they were joined by Situationist students, and *Noir et rouge* claimed that 'Cohn-Benditism' was none other than its own brand of 'anarcho-Marxism.'[4]

A fourth source of the ideas of 1968 can be traced back to the Marxist thought of Louis Althusser. A teacher at the École Normale Supérieure and member of the PCF, he devoted his seminar and his publications to a 'scientific' rereading of Marx, in particular of *Capital*. This was partly to attack the humanistic interpretation of Henri Lefebvre in defence of Marxist orthodoxy and partly, by demonstrating the Leninist doctrine that only intellectuals fully understood the scientific theory of socialism, to carve out a leading role for intellectuals like himself in the party. Ironically, one of his main theses, set out in *Pour Marx* (1965), turned out to inspire in his students heretical and anti-Leninist views. For his assertion that ideology and politics were not entirely determined by the economic substructure of forces and relations of production but were relatively autonomous seemed both to reflect and to legitimate Mao Tse-Tung's Cultural Revolution. Mao, who was trying to rival the Soviet Union for leadership of the Communist world, was also a proponent of the mass struggle against conservative elements in the party bureaucracy, quite the opposite view from the Leninist one that the party always knew best. Maoism spread among Althusser's pupils as a doctrine of mass struggle opposed to party bureaucracy, and of Third World revolution against the imperialism both of the United States and of the Soviet Union. While Althusser remained in the party, his Maoist pupils were expelled in 1966. One of them, Régis Debray, went to Cuba to interview Fidel Castro and met Che Guevara in Bolivia in March 1967. He publicized the Latin American dimension of opposition to American imperialism, and demonstrated that the mass struggle, making a nonsense of the idea of a Leninist party, there took the form of guerilla war in the countryside. Debray himself became a centre of attention later that year after Che was killed and he himself was put on trial by the Bolivian authorities, later to be released, ironically, thanks to the intervention of de Gaulle.

The mass struggle against imperialism in the Third World context was even more dramatic in Vietnam. Intellectuals including Sartre and students petitioned and demonstrated against American carpet-bombing increasingly after 1965. A network of Vietnam committees to campaign against the war was set up by Maoist students. Not only was support for Third World revolutions an important ingredient of the movement of 1968, but Maoists were a key element in the 'going to the people' movement which linked students and workers. They both won the sympathy of Sartre, who denounced the Communist party's fear of revolution in 1968, and ransacked the offices of Althusser, who remained loyal to the party line.

The gurus of structuralism

May 1968 was a carnivalesque inversion of all rules and norms imposed by society, and that included the doctrine of structuralism that was dominant in many universities in the 1960s. Structuralism was developed not by freelance intellectuals like Sartre but by ambitious academics who, finding the disciplines of the traditional education system too limiting for them, followed chequered careers before securing professorships at élite research institutions such as the École Pratique des Hautes Études and the Collège de France. Whereas for Sartre philosophy was the queen of sciences, the structuralists made their mark in, and indeed developed, the emerging social sciences of linguistics, sociology, anthropology, and psychology. Roland Barthes, who took a degree in classics in 1939 and pioneered the structuralist criticism of literature, was held back in his career by tuberculosis and divided his time between teaching abroad in Romania and Egypt, publishing, and spells in sanatoria before his election to the École Pratique in 1962. There he joined Claude Lévi-Strauss, who had taken degrees in law and philosophy and taught in provincial lycées before accepting a professorship in sociology at the university of São Paolo in 1934, which enabled him to do anthropological fieldwork in the interior of Brazil. He spent the war years in New York studying linguistics, the structuralist methods of which he then applied to anthropology, returning to the École Pratique in 1948. Jacques Lacan, who trained as a psychiatrist in the Paris Medical Faculty, developed a structuralist reading of psychoanalysis which entailed his expulsion from the International Psychoanalytical Association in 1953. In 1963 his seminar moved to the École Normale Supérieure, where he was close to Althusser, and he became a lecturer at the École Pratique. Michel Foucault, somewhat younger than the others and a pupil of Althusser, taught philosophy in Sweden, Poland, and Germany before securing a chair in philosophy at Clermont-Ferrand. But his real interest was in the history of science and medicine, particularly the history of madness, and it was his *The Order of Things*, an intellectual history, published in 1966, which resulted in his promotion to the new university of Vincennes.

Whereas Sartre believed that language was clear and transparent, referring unambiguously to the real world, the structuralists argued that language was opaque, independent of the real world, arbitrarily related to it, itself defining reality rather than merely reflecting it. Moreover, they saw it as a closed system governed by its own elaborate rules, which it was the task of the structuralist to lay bare. Structuralists had no time for the creative individual, and saw man not as the creator of language and other cultural forms but as created by them; commonly they would speak of the death of the author or the death of the subject. They were resolutely opposed to history, having no interest in the origins of language or other cultural forms but only in their internal structures. Lastly, they believed that all cultural forms including myths and the unconscious, were structured like language, and that they made use of such devices as metaphor and metonym.

In 1963 Barthes published a book on Racine that was a direct attack on the work of the eminent Sorbonne professor Raymond Picard. He argued that knowledge of the life and times of Racine was of no use in understanding his work, and that the literary critic must focus uniquely on the text itself, laying bare its internal oppositions and the way it incorporated and reworked other texts and myths. Further, he claimed

154 *Cultural Revolutions*

that there was no one 'correct' reading of a text but a plurality of meanings, with no hidden kernel of truth. This structuralist analysis was already at work in other disciplines. In his *Structural Anthropology*, published in 1958, Lévi-Strauss attacked functional anthropology, which held that all rituals and myths fulfilled a practical function. He argued that the totemic classification of primitive peoples that related human groups to plant and animal groups was not designed to prevent incest between different human groups, but was simply a language for apprehending and structuring the world. Similarly, he demonstrated that myths all had a similar structure and explored a finite number of relationships concerning blood relations (the Oedipal myth) and the emergence of man from Nature. The achievement of Lacan, with his *Écrits* of 1966, was to attack the American view of the free and autonomous ego, arguing that it was split and decentred by the paternal ban on incest with the mother and the repression of these unthinkable desires into the unconscious as the Oedipus complex was assimilated. The unconscious, Lacan continued, is structured like language, accessible only by language, and makes use of puns, rhymes, and word-associations. In *The Order of Things*, finally, Foucault abandoned the study of the history of ideas by individual author and individual text, instead seeing both conditioned by a dominant discourse that straddled disciplines and ordered the world in different ways in different periods. This ordering was conditioned by language, so that in the Renaissance thinkers structured the world according to resemblances between phenomena, in the classical age they looked to tabulate differences, while in the nineteenth century they were concerned by one thing following another.

This structuralist thought, squeezing out the creative individual and the outside world, was seen as arid and deterministic by the students of May 1968, who believed in asserting will, desire, and imagination. 'Down with structuralism!' and 'Power to the imagination!' were among their slogans. There were already some signs of change with a spate of publications in 1967 by the 37-year-old Jacques Derrida, a *pied-noir* who had studied at Harvard and taught at the Sorbonne and the École Normale Supérieure. Whereas the task of structuralism had been to discover the internal oppositions that propped up texts, Derrida sought to discover the gaps, contradictions, and impasses in texts which undermined their coherence, allowing them to be torn apart and rebuilt with a totally different meaning.

This deconstruction opened the way to an iconoclastic attack on canons, myths, and ideologies, but the failure of the movement of 1968 revealed that the rules and norms imposed by society were not just linguistic, and that the structures of power were manifold and complex. Intellectuals in the 1970s thus devoted themselves to examining the structures of power, the better to deal with them. In 1975 the philosopher Gilles Deleuze and the Lacanian psychoanalyst Félix Guattari published *Capitalism and Schizophrenia. The Anti-Oedipus*. This examined how man, a creature of desire, was forced by capitalism to squeeze his desires into work for achievement and private property, and by acceptance of the Oedipus complex to channel them into family life. Schizophrenics, who refused to assimilate these norms, were defined as mad by psychiatrists on behalf of society, but in fact, argued the authors, they alone understood revolution. With *Discipline and Punish*, published in 1975, Michel Foucault switched his interests from knowledge to power. He argued that relations of power were everywhere in society, and examined how prisons, asylums, workhouses, factories, barracks, and schools had emerged in the nineteenth century, developing techniques of surveillance and imposing norms the better to control individuals. He also began a *History of Sexuality*, examining how norms of sexual behaviour had been imposed and other practices such as homosexuality labelled as deviant and punished. He himself became more militant, campaigning for prison reform and gay rights. These new studies of power indicated that the old Marxist doctrine that liberation would come from proletarian revolution was redundant. Just as power was ubiquitous in society and repression many-sided, so the struggle against it would have to tackle it from all angles, including ecologism, regionalism, anti-racism, feminism, and the gay movement.

The new philosophers

For the generation of intellectuals who began their public careers in the 1970s, Marxism was not only redundant but evil. Two of them, André Glucksman and Bernard-Henri Lévy, were pupils of Althusser who had flirted with Maoism and taken part in the events of 1968, but who reacted to the failure of revolution not by looking for alternative modes of resistance but by becoming disillusioned with revolution itself. For them the event of the decade was the publication in French in 1974 of Solzhenitsyn's *Gulag Archipelago*. The image of the Soviet Union, long tarnished by purges and invasions, was now redefined as one of labour camps, psychiatric hospitals, the stifling of dissidence and, indeed, of intellectual life. The term 'totalitarian', once only applied to Nazi Germany, now came to characterize the Soviet Union.

For Glucksman in *The Cook and the Cannibal* (1975) and Lévy in *Barbarism with a Human Face* (1977), this totalitarianism was not the product of Asiatic despotism but was inherent in Marxist thought itself. It opened the way to an attack on socialism as a pack of lies, leading to totalitarianism, on the French Left as tainted with fascism and ant-Semitism, and on the French Revolution itself. In 1978 the revisionist historian François Furet argued that the ideology of 1789 led straight to the dictatorship and terror of 1793, just as the ideology of Marxism led directly to the dictatorship and terror of the Soviet state. An Anti-Totalitarian Front was set up by intellectuals who included François Furet, Raymond Aron, and Lefort and Castoriadis, launched on a new career after the demise of *Socialisme ou barbarie* ten years before. The language of liberalism—political rights and pluralism—was developed in opposition to the Marxist and totalitarian menace. Alexis de Tocqueville was rediscovered and the United States of America, long feared and hated by French intellectuals, was now hailed as the promised land. The thinkers of the New Right, led by the philosopher Alain de Benoist and his Groupement de Recherche et d'Études pour la Civilisation Européenne (European Civilization Research and Study Group, or GRECE), with their shop-window in *Figaro-Magazine*, tried to seduce these liberal intellectuals. Though they could agree on anti-Marxism, however, Alain de Benoist was as opposed to liberalism as he was to Marxism.

The result of this was that when the Left finally came to power in 1981 it did so, unlike at the time of the Dreyfus Affair, Popular Front, or Liberation, without the enthusiastic support of the intellectual community. They could not trust a Socialist party in alliance with the Communists, 'Stalinism with a human face', as Lévy had called it, and obliged to follow a foreign policy dictated by Moscow. This, they claimed, became clear when martial law was imposed in Poland in December 1981 and the French government said that it would take no action since this was an internal matter for the Poles. A wave of petitions argued that the coup in Poland was no different from that in Spain in 1936 or Hungary in 1956. In an important article in *Le Monde* in July 1983, the government spokesman Max Gallo bemoaned 'the silence of the intellectuals of the Left' when it was under attack from the New Right and needed support for its programme of economic and social modernization.[5] But it ignored the fact that, with few exceptions, French intellectuals were no longer on the Left.

The end of the French intellectual?

The demise of the left-wing intellectual was compounded by another phenomenon, the decline of the intellectual as such, or at least a transformation into something quite different from the days of Sartre. Sartre himself died in 1980, Raymond Aron in 1983, Simone de Beauvoir in 1984. But the same years saw the deaths of the following generation of intellectuals, Barthes in 1980, killed in an accident outside the Collège de France, Lacan in 1981, Foucault of AIDS in 1984, Althusser after years of mental illness in 1990. Normally, the French intellectual community would renew itself generation by generation, but arguably it was no longer a community in the 1990s and the best-known intellectual was the pretentious lightweight Bernard-Henri Lévy.

A number of factors may help to explain the decline of the French intellectual. One was the sharper definition of professions and their field of expertise on the one hand, and the public deference to proven experts on the other. Experts did not hesitate to express opinions in fields outside their own, and even to participate in the production of influential periodicals. A grounding in some sort of expertise nevertheless legitimated their wider view. Freelance intellectuals like Bernard-Henri Lévy, imitating Sartre with an editorial position at Grasset and an opinion on everything from women to Bosnia, no longer commanded the respect accorded to Sartre but were generally regarded as superficial.

The intellectual was also transformed by the growth of mass media. Intellectuals, of course, always required a public and therefore needed to manipulate the media of the day. But whereas Sartre and his contemporaries had editorial control of their reviews, elaborated coherent schools of thought, and sought to influence the few who mattered, the mass media and the mediacrats who controlled them made their decisions in the light of sales and ratings. They decided which books were promoted and which authors selected to appear on Bernard Pivot's *Apostrophes*, an influential discussion programme which ran from 1975 to 1990. They were interested not in works but in celebrities, not in schools of thought but in sound-bites. Those intellectuals who fitted the bill were the likes of Bernard-Henri Lévy, with his romantic coiffure and Yves Saint-Laurent silk shirts, marrying the glamorous film star Arielle Dombasle before the cameras in 1993, no longer describing himself as a philosopher but as a writer, and quintessentially a media personality, lovingly exposed to every medium.

A final explanation for the decline of the intellectual is the transformation of the public. Intellectuals ceased to wield the same moral authority over the public. As a result of the democratization of education, the public held intellectuals less in awe. At the same time, however, the public became less politically aware, trained in school as cogs for the economy rather than as citizens, more concerned in the 1980s with an agenda of private happiness than with one of public responsibility. The post-modern environment overturned old hierarchies, saw all art as commodities and all commodities as art, considered Adidas to be of equal value to Apollinaire, and a designer, comedian, rock star, or footballer to be as worthy of attention as an intellectual. The philosopher Alain Finkielkraut put it even more radically. 'Non-thought, to be sure, has always coexisted with the life of the mind,' he said, 'but it is the first time in European history that this non-thought has donned the same label and enjoyed the same status as thought itself.'[6]

The culture industry

The development of mass culture—or of non-culture—was common to all Western societies after the war. Cultural products were manufactured on a mass scale, marketed by advertising, made ever more accessible by revolutions in technology. Although in France mass culture took hold less rapidly than in the United States, Great Britain, and West Germany, the threat that it posed was felt especially keenly in France. The French prided themselves on their intellectuals, and their authority, as we have seen, was undermined by the mass media. The French saw themselves as the bearers of artistic taste and high culture, and these were under attack from the standardized and mediocre products of the entertainment industry. Above all, mass culture was seen to be a vehicle of American imperialism, threatening the French song, the French film, and French literature with extinction. Whether these fears were justified in practice will now be examined.

Cinema and television

After the war the cinema enjoyed an immense popularity in France. Both in 1947 and 1957 over 400 million cinema tickets were sold. Although the cinema is often regarded as a medium in which the French excel, its popularity was somewhat modest compared with the craze in other countries. Over 4,000 million tickets were sold in the USA in 1945, 1,600 million in Great Britain in 1946, and 800 million both in West Germany and in Italy in 1955. Cinema audiences were predominantly urban and young, and the French population was neither very urban or very young immediately after the war; 40 per cent of its inhabitants in 1958 never went to the cinema. When the French did go to the cinema, they tended to watch American rather than

French films. The price of American aid to rebuild France was the opening of its markets to American exports, including Hollywood films.[7] Moreover, colourful and spectacular films provided a welcome means of escape from the austerities of the post-war era.

The French cinema really took off after 1959 with the New Wave. Directors such as Claude Chabrol, Jean-Luc Godard, and François Truffaut, who had made their debut as critics for the *Cahiers du cinéma* and were keen to escape the constrictions imposed by American production companies, formed their own companies to make films on a shoestring, and pioneered a new kind of filming, with a freely held camera, few actors, few locations, and themes from everyday life. Ironically, the 1960s, the decade of the New Wave, saw the number of cinema tickets sold slump from 370 to 175 million. Not all the blame can be laid at the door of the somewhat intellectual style of the New Wave, for the 1960s also saw a boom in television ownership. After 1975, agreements between television companies and the film industry secured finance from television for the production of films, a quota of 50 per cent for films of French origin, and scheduling to limit the televising of films at weekends, prime time for cinema-going. Cinema-going recovered in the early 1980s, to 200 million tickets a year, but the revival did not favour the French film. Between 1979 and 1989 its share of the market fell from 50 to 34 per cent, while that of American films rose from 29 to 55 per cent. The French government was obliged to mobilize the European commission to defend the French film, securing the exemption of cultural products from the GATT agreement on free trade in 1993. Meanwhile the profile of cinema-goers remained very distinct: in 1989 73 per cent of them were aged under 35, 70 per cent lived in towns of over 100,000 inhabitants, and 44 per cent had passed through higher education. At this point cinemas had to compete with the advent of the video. Videos came late to France, owned by only 31 per cent of households in 1990 compared to 66 per cent in Great Britain. But in 1992 half of French households had videos, and in that year only 115 million cinema tickets were sold.

Television, the greatest threat to the cinema, had itself come late to France. In 1963 there were only 3.5 million sets in France, as against 7.9 million in West Germany and 12.5 million in Great Britain. Britain launched a second channel in 1953, France in 1964, while retaining the state monopoly. But the boom was rapid. The proportion of French households with television sets jumped from 15 per cent in 1960 to 70 per cent in 1970, then increased steadily to 90 per cent in 1980 and 95 per cent in 1990. Whether television would be the friend or foe of culture was, of course, the central question. André Malraux, the minister of culture, observed that more people would see a Racine play in one night on television than had seen it in the theatre across all the intervening centuries. He would have loved to have secured control of television, but it remained the instrument of the ministry of information. Unfortunately the most popular television programmes were game shows, such as *La Tête et les jambes*, watched by 3–5 million viewers around 1960 and reducing theatre and cinema audiences by 20 per cent on a Thursday night, together with soap operas, which took off in the 1960s, and above all films. A survey of 1982 showed that the most avid watchers of television were the over-60s and those with only an elementary education, and that farmers, workers, and white-collar workers wanted only entertainment from their viewing. Greater competition was introduced among the channels in 1974, and in 1982 the Socialist government declared that 'audiovisual communication is free'. The first private (subscription) channel, Canal Plus, was set up in 1984, a fifth and sixth private channels (the fifth owned by Silvio Berlusconi) followed in 1985 and 1986, and TF1 was privatized by the Chirac government in 1987, acquired by the Bouygues group, outbidding Hachette. Francis Bouygues claimed to have no interest in culture, and in 1987 all the private television channels exceeded the quota of American films to which they were committed. Only the fact that the French were again backward in acquiring the latest technology—only 3.3 per cent of households having cable television and only 2.5 per cent satellite dishes in 1991—saved them from bombardment with a surfeit of American films by the international cable and satellite companies.

The written word

One of the phenomena of mass culture is said to be the displacement of the written word by predominantly visual means of communication. A survey of the readership of books, newspapers, and magazines in France after the war indicates the need for a more nuanced analysis. The production of books was, after all, one sector of the culture industry. Hachette launched the paperback in 1953 and monopolized it until the end of the 1960s; between then and 1981, paperback sales doubled from 50 to 100 million a year. The classic bookshop was forced to compete with the discount supermarket chain FNAC, which set up its first store in Paris in 1974 and was selling a third of all books by 1980. A medley of prizes modelled on the Prix Goncourt and the appearance of authors on *Apostrophes* was used to promote books.

Whether these developments increased the reading public is another matter. In 1960 about 50 per cent of the population did not read books. This was reduced to 30 per cent in 1973, but was still 25 per cent in 1988. The paperback did not so much democratize reading as make reading cheaper and easier for those who read already. Only 3 per cent of French people used public libraries in 1969, compared to 30 per cent of the British, and this figure grew only slowly to 10 per cent in 1980 and 16 per cent in 1982. The profile of the reading public was somewhat modified between 1973 and 1988. Women overtook men, older people read more, and younger people less. The *bande dessinée* or comic strip, which sold 20 million albums in 1980 and 30 million in 1990 and which might have served as a lifeline for the semi-literate, in fact served no such purpose. Designed for adults as well as students after 1959, it was read by those who were readers of other books also, often having university degrees, with medical students notoriously avid consumers. Far from being pulp fiction, the BD acquired cult status, notably among the generation of '68, sanctioned by the *Cahiers de la bande dessinée* after 1969 and the annual BD fairs at Angoulême from 1974 and Blois after 1984. Lastly, while science fiction in France never ceased to be a colony of American science fiction, the BD was defended as a particularly French (if not Belgian) icon.

One clear casualty in the post-war era was the national daily newspaper. It flourished at the Liberation, with 28 titles having a combined circulation of 6 million in 1946, but by 1953 there were only 12 titles and a circulation of 3.5 million copies. Competition came from radio news, as the number of radio sets doubled from 5.3 to 10.1 million between 1945 and 1957. Thereafter the position of the national dailies remained fairly constant, with 11 titles and 2.9 million copies in 1988. In contrast to this the regional and local daily press flourished, with a constant circulation of about 7 million copies since 1950. Whereas in 1988 *Le Monde* sold 387,000 copies and *Le Figaro* 432,000, *Ouest-France* sold 765,000. The local press carried the local gossip that brought together the local community, and its most assiduous readers were older, less educated people in the countryside who generally read nothing except the local paper.

As the national dailies declined, so the magazines took off. Illustrated magazines like *Paris Match* reached a peak of 1.5 million in 1960, but suffered in competition with television. Successful magazines addressed themselves not to the market generally but to a specific clientele as it took shape as a result of economic modernization, social change, or the multiplication of leisure activities. Thus news magazines modelled on the American *Time*, such as *L'Express*, launched in 1953, and *Le Nouvel Observateur*, were addressed to busy managers and professionals who did not have time to read a newspaper every day, and collectively sold 1.4 million copies in 1988. Women's magazines in the same year included 57 titles with a combined circulation of over 17 million copies; *Femme actuelle*, founded in 1984, alone sold nearly 2 million copies. The sports press held up well, *L'Équipe* selling 230,000 daily and the weekly *Équipe-magazine* 258,000 in 1988, while specialized magazines such as the monthly *Auto-Moto* sold 330,000. For the great French public, however, the best-sellers were the TV magazines. *Télé-7 jours*, launched by Hachette in 1960, sold 2 million copies in 1965 and 3 million in 1988. As if to prove that the market was not saturated, *TV Magazine* and *TV Hebdo*, both launched in 1986, respectively sold 3.6 and 1.6 million copies in 1989. Of the reading public, 15 per cent never read anything but TV magazines.

Music

One of the reasons given for the decline of reading among young people in the 1980s was its displacement by listening to music. The music industry, above all, was revolutionized by technological developments. The 1960s were the era of the LP and the transistor radio, the 1970s that of the audio cassette, while in 1988 56 per cent of French people had a stereo system (compared to 29 per cent in 1981), 31 per cent had a walkman, and 11 per cent compact discs. The availability of recorded music may have driven audiences away from live concerts, and certainly the proportion of those who went to concerts was much smaller, but the proportion of French people who had been to a rock or jazz concert in the previous twelve months doubled from 7 per cent in 1973 to 13 per cent in 1988.

The music industry was international, and even in the sphere of popular music France had a cultural heritage to defend. One reason for the slowness of the rock revolution to hit France was the strength of the tradition of the French *chanson*, which prided itself on being poetry set to music and was beloved of the cabarets of Paris. After the generation of Maurice Chevalier, Édith Piaf, and Tino Rossi a new team took up the baton in the early 1950s, notably Georges Brassens, Jacques Brel, and Léo Ferré. Though they graduated from the cabaret to the concert hall they prided themselves on their literary merit: Brassens was published by Seghers in 1963 and awarded the Académie Française's poetry prize in 1967. American-style rock music made its impact in France in the person of the 16-year-old Jean-Philippe Smet, alias Johnny Halliday, in 1959, followed by the Bulgarian-born Sylvie Vartan (whom Halliday later married) in 1961 and Françoise Hardy in 1962. Denounced as idiotic, illiterate 'yé-yé' by the apologists of the *chanson*, it was taken up by *Salut les copains!*, a Europe-1 radio show in 1959, and a fanzine in 1962. On 22 June 1963, the eve of the departure of the Tour de France, it promoted a rock concert on the place de la Nation, the bill headed by Johnny Halliday, which was attended by a crowd of young people estimated at between 50,000 and 150,000. Halliday's reputation rose and fell, but though rarely away from the glare of publicity he and his fellow rock stars always faced an uphill struggle against the *chanson*. Brel died in 1978, Brassens in 1981, and Ferré in 1993 but the French song was revived by new talents such as Serge Gainsbourg, J. J. Goldman, and Renaud, then by Enzo Enzo and Kent. A survey of 1988 showed that while young people aged 15–19 listened overwhelmingly to English and American songs and current French hits, a third of French people aged 20–4, half aged 25–44, and nearly two-thirds of those over 45 listened regularly to the French *chanson*. The French government was also happy to respond to pressure from the French music business, and Jacques Toubon, minister of culture in the Balladur government, sponsored a law to give French songs 40 per cent of time devoted to popular music on the airwaves from 1996.

Sport

Sport, which may be considered a branch of the culture and leisure industry, was increasingly colonized by the mass media and advertising. Sport sold newspapers and magazines and drove up television ratings. Television exposure attracted advertising to the media, and television rights and advertising returned finance to the most televisual sports, football, tennis, motor-racing, and cycling. Increasingly, sport was watched not from the stands or terraces but from the armchair. The proportion of the population that had been to at least five sporting events in the previous year dropped from 17 per cent in 1967 to 9 per cent in 1987.

At the same time that it multiplied the number of armchair specta-tors, however, television also promoted active participation in sport. Sports such as tennis were no longer confined to the few in private clubs but attracted mass participation. A census of 1988 counted 1.8 million footballers, 1.4 million tennis-players, and nearly a million skiers among the paid-up members of sporting federations, not counting casual players. For all the colonization by the media, however, French sport retained a certain Gallic individuality. The Tour de France, sponsored from the 1970s not by the depressed cycle industry but by Fiat, Coca-Cola, and the Crédit Lyonnais, and watched by 150 million viewers in 1976, remained a celebration of French regional vitality and was the only international competition to take place on the doorstep. Horse-racing, once the aristocratic sport *par excellence*, was popularized by the invention of the *tiercé* in 1954, which extended betting beyond the five or six major races of the year. There were 960 agencies of the Paris Mutuel Urbain (PMU) in 1954, 5,600 in 1978. Located not beyond public view but in the cafés, bars, and bistros of France, attracting 7 or 8 million punters, mainly from the urban working class and petty bourgeoisie, every Sunday morning, they animated the life of the *quartier* and acted as a focus for the community far more than the churches that the French were deserting in droves.

The old cultural policy

Despite the barrier of the French language and the relative slowness of successive technological revolutions to take hold on France, French governments were extremely concerned by the spread of Anglo-Saxon mass culture and by what they saw as threats to the arts and high culture. More than other governments, excepting that of the Soviet Union, they developed cultural policies designed to defend culture and French culture in particular. What that culture might be and how it might be spread, however, were both extremely problematic.

After the Liberation there was a sense that the masses had acquired time for leisure and also had a right to enjoy culture, but that the innocent enjoyments of rural communities had been destroyed by urbanization, and that culture itself was being blotted out on high-rise estates, by the consumer society, and by the mass media. In 1945 Jean Guéhenno, the self-taught son of a shoemaker who became a philosophy teacher in a Parisian lycée and militant of the Popular Front, was appointed Directeur de l'Éducation Populaire. His right hand as Inspecteur Général de la Jeunesse et de l'Éducation Populaire was Joffre Dumazedier, the son of a mason, a trade unionist who had been active both in the Popular Front and in Vichy's École des Cadres at Uriage, which, far from the centre of power, had been committed to the cultural and spiritual regeneration of France. The project of Guéhenno, Dumazedier, and their imitators in the Popular Culture movement was to generalize from their own experience, to turn the right of the working class to high culture into a genuine access, to bring people out of their flats to participate in arts and crafts, play sports, join libraries, go on organized visits to the theatre, and to teach them how to deal critically with cinema and later television through 'ciné-clubs' set up after 1946 and 'télé-clubs' after 1950.

The most notable landmark of this period was the Théâtre National Populaire, established at the Palais de Chaillot in 1951 and directed until 1963 by Jean Vilar. Designed to bring the great works of the theatre to the masses, and working together with the *comités d'entreprise* (works councils) of large firms, it first organized trips to the town halls of the 'red belt' around Paris, then (given logistical and budgetary difficulties) arranged special 'works outings' to Chaillot itself. The diet of Shakespeare, Racine, and Brecht—with no new writer in sight— however, was far more popular with students and young people than with workers. To bring his message to the provinces, Vilar promoted the annual Avignon festival which began in 1947. Performing the classics on a vast stage in the Palais des Papes, subsidized by the municipality, the festival was popular mainly in the sense that it attracted large numbers of tourists.

The state's view of culture at this point was that it was a sacred heritage, to be transmitted intact to every class and each generation. None defended this position more firmly or more vigorously than André Malraux, appointed minister of state for cultural affairs by de Gaulle in 1959. His mission as he saw it was to 'make accessible to the greatest number of French people the greatest works of humanity, beginning with those of France, to achieve the widest audience for our cultural heritage, and to promote the creation of works of art and the mind that will enrich it.'[8] In the event he was more interested in showing off the classical heritage than in sponsoring new creations. The Louvre was more important to him than the Musée d'Art Moderne, which in 1959 had no works by Klee, Munch, Magritte, or Sutherland and seventeen Picassos only because the artist had donated 16 of them. In reply to the challenge of the New York School of painting, headed by Jackson Pollock, which had a successful exhibition at the Musée d'Art Moderne in 1956, Malraux sent the *Mona Lisa* to Washington in 1963 and the *Venus de Milo* to Tokyo in 1964.

Alongside the museums he planned a network of Maisons de la Culture, to be built in the context of the Fourth Plan and to serve as cathedrals of culture in urban centres. The first Maison de la Culture opened at Le Havre in 1961 and others rapidly followed. They were designed as powerful weapons for the democratization of culture, especially of the theatre. Very soon, however, difficulties emerged. The provincial bourgeoisie preferred to see plays by Molière and Labiche rather than Brecht or Beckett. The municipalities who represented them refused to contribute their share of the budget, and directors of the Maisons de la Culture met in anger at Villeurbanne in May–June 1968 to demand greater subsidies. The working classes stayed away—they accounted for 4 per cent of the 30,000 members of the Grenoble Maison de la Culture which opened in 1968—and students were keener to experiment with their own theatre. Jack Lang, who founded the student theatre of Nancy in the late 1950s, launched the international festival of student theatre at Nancy in 1963. Its plays were experimental, encouraged collective creation, took to the streets, attracted exciting international troupes like the New York-based Bread and Puppet Theatre, and engaged in political debate. The events of May 1968 in some sense began at Nancy, while in July 1968 students spilling out from the May events in Paris hiked down to Avignon to disrupt Vilar's festival, denouncing him for accepting the commercial criteria laid down by the municipality and for 'having admirably played the repressive and authoritarian role assigned to him by the ruling class.'[9]

The new cultural policy

The events of 1968 demonstrated the failure of the state's cultural policy. It was prescriptive, and sought to impose a model of French culture from the top down. It was monumental, preserved in temples of beauty which made a nonsense of the mission of democratization. It invited the passive worship of masterpieces of creation, not participation in the creative process. If a way were to be found out of the impasse, cultural policy would have to be far more receptive to the local and popular manifestations of culture, conceived in its broadest sense, for example articulating museums around local industry, crafts, history, and art. It would have to dismantle the whole notion of the museum, which repelled more than it attracted, turning it inside out in order to make it accessible. And it would have to give value to the creative act and elicit the creativity of every individual rather than reserving it for the works of artists which conformed to some established canon of greatness.

One response to this challenge was the Théâtre du Soleil of Ariane Mnouchkine. Having performed Arnold Wesker's *The Kitchen* in the occupied factories in 1968, she staged *1789* in the Cartoucherie of Vincennes in 1971. The creation was a collective act, the style was that of a fairground or circus, and the action enclosed and involved the audience who stood, as the crowd in the French Revolution. A quarter of a million people flocked to see the show. Meanwhile a design competition was held for a new Musée d'Art Moderne, combined with a library and information centre using the latest technology. The design of Piano and Rogers, chosen from 681 entries, became the Beaubourg or Pompidou Centre, opened in 1977. It was supposed to be the antithesis of a museum, turned inside out with its guts showing, the sight of visitors floating up its perspex escalators attracting more visitors, the emphasis on confrontation and debate with art and open access to the resources of book, film, and video libraries. The Beaubourg was certainly popular. Planned to receive 7,000 visitors a day, it attracted 25,000 a day, the same as the Eiffel Tower. Whether the result was the democratization of culture was another matter. Tourists climbed it for the view, students plundered the libraries, others responded to the challenge to 'bend Beaubourg', the metal frame of which was said to give way under a critical mass of 30,000. 'Mass culture is being destroyed by the masses themselves,' bemoaned one critic.[10]

When Jack Lang became minister of culture in 1981, the cultural revolutionary was given the power to effect a cultural revolution. The indefatigable Lang stimulated a frenzy of creativity. While the intellectual establishment snubbed the advances of the Socialists, the artistic proletariat was fêted and subsidized. A Fête de la Musique was held each Midsummer's Day, 21 June, to incite musical talent. Anglo-Saxon music was denounced and Lang ostentatiously refused to open the festival of American film at Deauville. But in its drive to succeed, the Lang enterprise harnessed the twin motors of politics and the mass media, and art was placed at the service of power and publicity. As Jean-Louis David had been the pageant-master of the First Republic, so Jack Lang was the pageant-master of the Fifth. All Lang's projects were on the grandest scale, too many of them over-ambitious. The climax of the Bicentenary of the French Revolution on 14 July 1989 was a spectacular parade down the Champs-Élysées, complete with tam-tam drummers and London buses, watched by a million people. The new Opera at the Bastille, due to be ready for the Bicentenary, not opened till 1990, lost its musical director, Daniel Barenboim, had three general directors in as many years, and was cursed by disasters such as collapsing scenery. The coffers of state were opened in 1986 to create a wave of Zeniths—concert halls to seat up to 7,000 people—in all the major cities. By 1993 Zeniths had opened at Paris, Montpellier, Pau, Toulon, Nancy, Caen, and Marseille, but Lyon could not afford one, having spent all its money renovating its opera house, Tours preferred a Palais des Congrès, and Rouen, Rennes, Bordeaux, Toulouse, Clermont-Ferrand, Saint-Étienne, Grenoble, and Nice were not interested.

The Lang cultural revolution, which lasted (with the exception of 1986–8) until 1993, was quite unprecedented. It could, however, be criticized on a number of grounds. First, the desire to communicate took precedence over the object communicated, and the masses had become so fond of mass culture that they threatened to destroy it. Jacques Toubon, Lang's successor, considered closing the crumbling Pompidou Centre in 1994, and instead launched a rebuilding job estimated to cost 800 million francs. Second, paradoxically, a survey of French cultural habits in 1988 showed that the vast majority of French people were non-users. Eighty-two per cent had never been to the opera, 76 per cent had never been to the ballet, 75 per cent never to a rock concert, 71 per cent never to a classical concert, 62 per cent never to an art gallery, and 55 per cent never to the theatre. Moreover, despite the revolutions in cultural policy, the improvement since 1973 had been minimal, 3 per cent on average at the most. Third, Malraux's idea that culture was the cultural heritage of France to begin with, then the rest of the world, was replaced by a totally elastic view of culture according to which every profession, social group, or community had its culture and Benetton was as good as Beethoven. No doubt this 'anything goes' attitude only reflected post-modernism, but many an eyebrow was raised when the French minister of culture decorated Sylvester Stallone for services to art. Lastly, subservience to the mass media undermined the defence of French culture and opened the floodgates to the American culture and leisure industry, symbolized by the opening of EuroDisney on the outskirts of Paris in 1992. To their great credit the French public did not stampede to the park, and within a year or two the whole project was at risk.

7

The Republic of the Centre

Whereas between 1945 and 1968 the French state had repeatedly been in crisis, after the fall of de Gaulle it enjoyed a period of unprecedented stability. The constitution of the Fifth Republic, which had been shaped by de Gaulle for his own purposes, came to be accepted by the Left as well as by the Right. A regime that was intended to be enfeoffed to the Gaullist majority adapted itself to the alternation of Right and Left in power, becoming a healthy pluralist democracy, and also survived the coexistence of Left and Right in power at the same time. The guerilla war of Left and Right, fought as if the French Revolution were still going on, became attenuated. People voted on the Right or on the Left but were governed from the Centre. Parties of the Left and Right were seldom able to govern alone, and sought to broaden the majorities on which they were based by *ouverture* or bridge-building to include parties of the Centre in government. Moreover, the alternation of Right and Left in power became less problematic as consensus was established under which governments of both persuasions accepted the mixed economy and a certain level of health and social security benefits as given, and not to be tampered with from one government to the next. The dominant ideologies were less statist Gaullism or statist socialism but a liberalism which had had a poor track record in France since the Revolution but which now penetrated the thinking of all the major parties.

There was, however, a reverse side to this story of political harmony. While the Republic of the Centre entrenched itself, opposition grew up at the extremes of politics to what was seen as politicians who swapped power between themselves, had the same bland liberal ideas, and were confined to the same narrow economic and social agenda. This opposition, which made itself felt in the 1980s, took the form on the one hand of the National Front and on the other of the Green parties. In addition, the rising trend of electoral abstention betrayed a crisis of representation, a growing disillusionment with the political class of professional politicians who were regarded as both incompetent and corrupt. This was manifested in a decreasing interest in the traditional politics of parties and trade unions. It did not necessarily signify the political illiteracy of French citizens but rather indicated that they were more demanding of their leaders, and more adapted to pursuing individual needs in the private sphere. They wanted the state off their backs and the enlargement of civil society in the sense of those institutions and practices that were not the responsibility of the state. If they were political, they pursued a new kind of politics, outside that of the conventional parties and trade unions. This took the form of enthusiasm for political mavericks who called the system into question, support for single-interest groups that mobilized for specific purposes, and involvement in 'Band-Aid' politics which interlocked with show business and the mass media.

The 'new society' of Chaban-Delmas

On 28 April 1969, immediately after de Gaulle had left office, Jacques Chaban-Delmas, speaker of the National Assembly, called a meeting of the 'barons' of the Gaullist party—Georges Pompidou, Michel Debré, Olivier Guichard, Jacques Foccart, and Roger Frey—at his official residence, the hôtel Lassay, to determine the succession and ensure the continuity of Gaullism. It was decided that Pompidou should inherit the presidency. Placing the emphasis on *'ouverture* in continuity', Pompidou had the support of the Gaullist UDR, the Independent Republicans of Giscard d'Estaing, who rallied at the last moment, and a fraction of the Centre Démocrate which called itself the Centre Démocratie et Progrès (Centre for Democracy and Progress, or CDP). The Left was in complete disarray, unable to find a candidate to represent it as a whole, and failed to get through to the run-off ballot on 15 May. Pompidou's main opponent was Alain Poher, the speaker of the Senate who headed the interim government after de Gaulle's departure. A former member of the MRP, Poher's main idea was that the president of the Republic should be an 'arbiter', as originally defined in the constitution, even the benign president of the Fourth Republic, not the leader of a party or 'guide'. But Pompidou, who defended the strong presidency that de Gaulle had left, routed Poher, receiving 58 per cent of the vote.

Pompidou took Chaban-Delmas as his first prime minister. The Gaullist barons took the top jobs in the government, but Jacques Duhamel of the CDP entered the cabinet in recognition of his party's support, while Giscard d'Estaing became finance minister after Pompidou failed to persuade the 78-year-old Antoine Pinay to accept the post. Chaban, however, saw eye to eye neither with Pompidou nor with the conservative Gaullists. A progressive Gaullist himself, formerly a Radical, he looked to detach the Socialists from the Communists with a view to a broad alliance for the parliamentary elections of 1973. He took as his advisers Simon Nora, who had served in the European Community and advised Mendès France, and Jacques Delors, who had come up through the MRP and Catholic trade union movement before working on the Plan. Conflating John F. Kennedy's New Frontier and Lyndon B. Johnson's Great Society, they came up with the formula of a New Society, on which Chaban gave a major speech to the National Assembly on 16 September 1969. He described France as a stalemate society, with a backward and fragile economy, tentacular bureaucratic state, and caste-ridden social structure, and proposed massive investment in infrastructure and training, greater autonomy for local government, universities, nationalized industries, and the broadcasting service (ORTF), and improvements in the minimum wage and social benefits.

Pompidou was infuriated not only by the tenor of the speech but by the fact that Chaban had not consulted him in advance. The barons, who tended to see Chaban as the spoiled child of Gaullism, were annoyed by the disdain in which he increasingly held the party and hated his ally, Giscard d'Estaing, as the Brutus who had stabbed the General in the back on 27 April. Pierre Juillet, Pompidou's personal adviser, was suspicious of Paris intellectuals and technocrats. Regarded as the voice of 'la France profonde' of the Massif Central (in his case the Creuse, in Pompidou's the Cantal) and of conservative Gaullism, he accused Chaban of 'bringing socialism into France', handing the television over to their opponents, and undermining the authority, of the state.[1] Juillet was the spider at the centre of the web that orchestrated the palace coup. Chaban was accused of tax evasion by *Le Canard enchaîné* in January 1972, and sacked on 5 July. He was replaced as prime minister by Pierre Messmer, who held much more strongly to the tenets of conservative Gaullism.

The 'advanced democracy' of Giscard d'Estaing

Georges Pompidou died in office in April 1974. Chaban-Delmas saw himself as in a strong position to take up the mantle, but despite his enduring good looks he was hampered by a metallic voice and snobbish accent. Messmer tried to stop him running, fearing that he would lose to Mitterrand, but Chaban persisted. Meanwhile the 48-year-old

Giscard was in a hurry to step over him on the way to the presidency, and secured the support of Jacques Chirac, a protégé of Pompidou who had risen to be minister of agriculture, then minister of the interior, under Messmer. Though there was no love lost between Giscard and Chirac, the one an Orleanist relishing the social and intellectual élite, the other a Bonapartist enjoying the company of the Corrèze peasants who had elected him since 1967, they concluded a marriage of convenience for the sake of achieving power. In the first round of the presidential election, on 5 May 1974, Giscard trounced Chaban, obtaining 33 per cent of the vote to Chaban's meagre 15, while Mitterrand streaked away with 43 per cent. But in the run-off, promising 'change without risk', Giscard just nipped home with 50.8 per cent of the vote.

Giscard had said as early as 1972 that he wished to govern France from the centre. His own power base, the Independent Republicans, was narrow, but he failed to take the opportunity to dissolve the National Assembly, which might have resulted in the election of a large centrist majority. As a result, he was obliged to make Jacques Chirac his prime minister as a reward for his bringing over the Gaullists, while offering portfolios to the leaders of the centrist parties. He won over the Centre Démocrate, making Jean Lecanuet minister of justice and Simone Veil minister of health, and brought in those Radicals under Jean-Jacques Servan-Schreiber who had refused to ally with the Socialists and Communists. JJSS only lasted a few weeks as minister of reforms, removed after he took part in an anti-nuclear protest, but was replaced as standard-bearer of the Radicals by Françoise Giroud, appointed secretary of state for women's affairs. Chirac complained that his ministry was beginning to look like a branch of *L'Express*. Finally, Giscard even hoped at some time in the future to attract Socialists who refused to go along with the Communist alliance. Giscard projected himself not only as a centrist but as liberal and modern. Initially he sought to demystify the presidency, attending his inauguration in a plain suit instead of formal dress and having himself invited to lunch in the homes of ordinary people. He had the 'Marseillaise' rewritten as a hymn rather than as a battle-song, reduced the voting age from 21 to 18, and extended women's rights by allowing Simone Veil to sponsor a bill conceding the right to abortion. He broke up the ORTF into seven competing broadcasting companies, made the Conseil Constitutionnel a genuine watchdog of the constitution instead of a poodle of the executive, and allowed Paris to elect its own mayor for the first time since the Revolution. At the same time, however, Giscard was vain, authoritarian, and jealous of the prerogatives of the presidency. He and his minister of the interior, Michel Poniatowski, a close personal friend, were ruthless in their attempt to crush the Corsican autonomist movement by force. He interfered not only in the composition of the government but in the daily running of its affairs, relying on Poniatowski to keep a close eye on Jacques Chirac. President and prime minister locked in an increasingly fierce struggle for power. Chirac used his position as prime minister to seize control of the UDR from the Gaullist barons in December 1974, and looked to use it as a vehicle to challenge for the presidency. In January 1976 Giscard and Poniatowski imposed a cabinet reshuffle on Chirac, promoting Jean Lecanuet to minister of state and bringing in Raymond Barre, an economics professor and former Brussels commissioner, as minister of foreign trade. Chirac's position became increasingly untenable, and after a row in the council of ministers on 25 August 1976 he stormed out and offered his resignation.

Out of power, Chirac renewed his own power base, founding a new Gaullist party, the Rassemblement pour la République (Union for the Republic, or RPR), deliberately modelled on de Gaulle's RPF, in December 1976, and having himself elected mayor of Paris in the municipal elections of March 1977. In response Giscard sought to refresh his modern, liberal centrist strategy. In place of Chirac, Raymond Barre was appointed prime minister. Not holding elective office himself, like Pompidou under de Gaulle, he was not in a position to challenge the president, whose bearing was becoming increasingly monarchical. It was intended that he should manage the economy as an expert and take the ideology out of politics. Giscard himself published his political testament, *Démocratie française*, which was somewhat like the manifesto for the New Society, seven years on. He rejoiced that the institutions of the Fifth Republic were no longer challenged, but looked forward to a time when archaic ideological battles would cease and the peaceful alternation of parties in power would be possible in France as it was in 'advanced democratic societies' such as the United States, Great Britain, or the Federal Republic of Germany. He preached pluralism, by which he meant the autonomy of organizations such as broadcasting bodies, parties, and trade unions, while affirming the need for a strong centralized state. He underlined the need for economic modernization but also the principle of social justice, by which he meant not equality or levelling but the equalization of opportunity and benefits to help those left behind in the race for modernization. Lastly, to ensure victory for the presidential majority in the 1978 parliamentary elections he relaunched his Independent Republicans as the Parti

Républicain (Republican Party), under Jean-Pierre Soisson, in May 1977, and in February 1978 put together a new coalition, the Union pour la Démocratie Française (Union for French Democracy, or UDF), headed by Jean Lecanuet. This was a cartel of the Parti Républicain, the Centre des Démocrates Sociaux, which healed the schism of 1969 of Lecanuet's Centre Démocrate and the CDP, right-wing Radicals, and the Parti Social Démocrate, grouping a small number of socialists who opposed union with the Communists. To set the tone for the elections, Giscard made a rousing speech at Verdun-sur-le-Doubs on 27 January 1978, warning that victory for the Social-ist–Communist alliance would be like the débâcle of 1940. He would never permit it to be said, as fleeing French soldiers had told him as a boy in 1940, 'we were conned'.[2]

The union of the Left and the 'break with capitalism'

Following the election of Georges Pompidou in 1969, the Left was in a disastrous position. The Communist party had betrayed the revolution

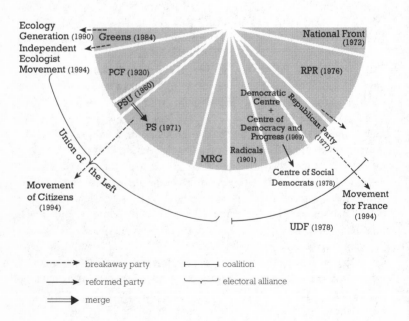

FIG. 8. Political Parties, 1969–1994.

of May 1968 and in August had expressed only 'surprise and disapproval' when the Soviet Union invaded Czechoslovakia. Although the PCF candidate, Jacques Duclos, had scored a respectable 22 per cent in the first round of the presidential elections, it was clear that the party would have to revise its Stalinist image if it wanted to re-enter the mainstream of French politics. The Central Committee's manifesto of Champigny-sur-Marne in December 1968 thus acknowledged that revolution was an end—the collectivization of economy and society—not a means—violence on the streets. Further, if the party were ever to regain power, it was imperative that it prevent the Socialists drifting into alliance with centrist parties and forge a union of the Left which it would naturally dominate, as the largest party of the Left since the Liberation. This became the strategy of the new secretary-general from 1972, Georges Marchais.

The Socialists, for their part, had come completely unstuck in the elections of 1969. Within the Fédération de la Gauche Démocratique et Socialiste (FGDS), formed by François Mitterrand for the 1967 parliamentary elections, the Convention des Institutions Républicaines (CIR) was no longer able to keep Mollet's SFIO in line. Whereas the CIR wanted to run a single candidate of the Left, that is, François Mitterrand, the SFIO refused to consult and unilaterally announced the candidature of Gaston Defferre. Defferre made a bid for the centre ground, taking Mendès France as a running mate, but scored a derisory 5 per cent. Socialists voted for Poher not only in the second round but also in the first. After this fiasco, the first priority was to forge a new disciplined Socialist party that would replace the moribund SFIO. The Parti Socialiste (Socialist Party, PS) was put together at Issy-les-Moulineaux in July 1969 and at Épinay in June 1971, and Mitterrand defeated Mollet to become first secretary with the help of Pierre Mauroy, who became mayor of Lille in 1973, and by a tactical alliance with Jean-Pierre Chevènement and his CERES group, who defined themselves as the guardians of the left-wing, Jacobin, Marxist strain of socialism that went back to Jules Guesde. The second priority was to foreswear alliances with parties of the centre in the tradition of the Third Force or Defferre's 'grand federation' and to build a union of the Left in partnership with the Communists. This would be not just an electoral pact, as in 1967, but a common programme of government. Hammered out between 1970 and June 1972, the PS accepted the dogma of the 'break with capitalism', involving a wave of nationalizations, while the PCF accepted that if the Left achieved power, it would not replace democracy by some dictatorship of the proletariat and go

quietly if at some future date it were voted out of office. For Mitterrand, however, the radicalization of the Parti Socialiste and the union
of the Left were designed not only to oust the Right but also to displace
the Communist party as the leading party of the Left. 'Our fundamental objective', he candidly told the Socialist International in Vienna on
27 June 1972, 'is to rebuild a great socialist party on the ground
occupied by the Communists in order to demonstrate that out of five
million Communist voters, three million can vote Socialist.'[3]

Socialists and Communists were thus engaged in a struggle for power
not only against the Right but against each other. Mitterrand was
happy to appropriate the Marxist critique of capitalism as a weapon
against the Right, but he was also keen to demonstrate the superiority
of Socialism to Communism by dint of its unreserved espousal of
liberty in the Republic. In *La Rose au poing*, published in 1973, he
argued that the Right represented monopoly capitalism and the dictatorship of the privileged, but he also quoted Léon Blum's denunciation
of the Communist dictatorship of the proletariat at the Congress of
Tours in 1920 and claimed that 'the socialists are the real heirs of the
Declaration of the Rights of Man and the Citizen.'[4] The very symbol
of the rose in the clenched fist was eloquent.

Initially the Communists retained the upper hand. The Communist
secretary-general Georges Marchais was visibly delighted when the
PCF won 21.5 per cent of the vote to the PS's 19 in the parliamentary
election of 1973. Equally, at the traditional garden party at the Élysée
on 14 July 1974, following Mitterrand's narrow defeat at the hands of
Giscard d'Estaing, Jacques Duclos was seen in excellent spirits sharing
petits fours with Jacques Chirac. But matters changed after the publication of Solzhenitsyn's *Gulag Archipelago* in France in 1974. The
Soviet Union was denounced as totalitarian by intellectuals and politicians and the PCF, one of the most Stalinist of Communist parties, was
caught in the crossfire. In vain the Communists tried to burnish their
image, formally abandoning the doctrine of the dictatorship of the
proletariat at their 22nd party congress in February 1976. The union of
the Left was broadened out to include the Mouvement des Radicaux
de Gauche (MRG). Mitterrand moved away from the Marxists in the
Parti Socialiste, and induced Michel Rocard, who preached *autogestion*,
decentralization, and the market economy, to abandon the PSU for the
PS in November 1974. The municipal elections were a heavy blow to
the Right and a great victory for the Left, who took 60 towns of over
30,000 inhabitants, so that they now controlled 160 out of 220 towns
of over 30,000. All the portents pointed to a triumph of the Left in the

parliamentary elections of 1978. At this point, however, the PCF leadership feared that the Socialists would overtake them as the largest party of the Left. First of all, suspecting that the Socialists would renege on the common programme, they demanded its renegotiation. Then, unilaterally, in September 1977, they ruptured the union of the Left.

The fears of the Communists were duly fulfilled. For the campaign of 1978 Mitterrand laid claim to the mantle of Jean Jaurès, like him a radical republican who had come late to socialism, who had wedded the SFIO to democracy, defended France as the cradle of liberty, and given his name to a thousand French streets. 'There is no reason, now, Jean Jaurès, to fear for liberty,' announced Mitterrand. 'The Socialist party is there, to guarantee it.'[5] Though the Left lost the elections in March–April 1978, the Socialist party won 23 per cent of the vote, the Communists 20 per cent. The Communist party was rocked by mutual recrimination. Its membership, younger, more female, and more middle class since 1968, blamed the defeat on the rupture of the union of the Left which the leadership had decided on without any consultation. Intellectuals such as Louis Althusser, forbidden from protesting in the party press, published their attacks in *Le Monde*. The PCF politburo made the gesture of meeting the party's intellectuals, but reacted by centralizing the party even more and squeezing out dissidents. On the eve of the Left's return to power the PCF was in crisis.

The modernization of Socialism

On 10 May 1981 there was delirium in the country as François Mitterrand, in his third bid for the presidency, defeated Giscard d'Estaing by 51.75 per cent of the vote to 48.25. In the first ballot Mitterrand had tied with Giscard on 26 per cent, with Chirac on 18 per cent and Marchais with a paltry 15 per cent. Whereas Chirac had backed Giscard in 1974, on this occasion he did not instruct his supporters to vote for Giscard in the second round, and 15 per cent of them voted for Mitterrand while another 15 per cent abstained. Georges Marchais told his supporters publicly to vote for Mitterrand. There is some evidence that militants were working behind the scenes to urge Communists to vote for Giscard, but the voters knew best and 92 per cent of Marchais's supporters transferred their votes to Mitterrand.

Mitterrand had long been a fierce critic of the constitution of the Fifth Republic, but once in power he did nothing to change it. 'France's institutions were not made for me,' he said, 'but they suit me well enough.'[6] The dissolution of the National Assembly had been considered an abuse of presidential power by the Left for over a century, but Mitterrand was happy to do it at once in order to secure a new majority for the Left. In the first round, on 14 June, the PS obtained 38 per cent of the vote and the PCF a mere 16 per cent. In the second round, the PS and their MRG allies secured an absolute majority of 285 seats, and the Communists slumped from 86 seats to 44. The government constituted under Pierre Mauroy included four Communist ministers, headed by Charles Fiterman at the ministry of transport, but this was not least to prevent the Communists setting up a 'dual power' based on the CGT against the government, and to retain their support with a view to the elections of 1986, when the Socialists might not be so popular.

At his inauguration, on 21 May, President Mitterrand declared that the government of the Left was in the tradition of the Popular Front and the Liberation, and would continue their work. He climbed the steps of the Panthéon alone to lay roses on the tombs of Jean Jaurès, Jean Moulin, and Victor Schoelcher, who had secured the abolition of slavery in the French colonies in 1848. He also sent his aide, Jacques Attali, to lay a wreath on the grave of Léon Blum at Jouy-en-Josas. The pace of reform was frenetic. The death penalty was abolished, as were military courts and the Cour de Sûreté de l'État, set up in 1963 and used most recently to try Corsican autonomists. Corsica was given a measure of autonomy and one of the most radical measures of decentralization since the Revolution, the law of 2 March 1982, was sponsored by the mayor of Marseille, Gaston Defferre. A whole battery of measures was taken at once to enhance social equality and reflate the economy by increasing consumption.[7] An Impôt sur les Grandes Fortunes or Tax on the Super-rich was introduced. The 'break with capitalism' took the form of the nationalization of two holding companies, nine industrial groups, and thirty-six private banks. At the same time the Loi Auroux of 4 August 1982 democratized industry by increasing the representation and power of workers.

At the Socialist party conference at Valence in October 1981, Paul Quilès, a deputy for Paris, played at being Robespierre and called for heads to roll. Far from imposing a reign of terror, however, the Socialist government undertook its own Thermidor. Though it enjoyed an absolute majority in the Assembly, other factors limited the extent of revolutionary change. The Conseil Constitutionnel, chaired by the old Gaullist Roger Frey, forced revisions to the nationalization law in the light of property rights enshrined in the Declaration of the Rights of Mán and the Citizen. Eighty per cent of heads of firms in the public sector lost their jobs in 1981, but their successors came from exactly the same background in the *grandes écoles* and *grands corps*. The strategy of reflation was called into question not only by galloping inflation, a balance-of-payments crisis, and budgetary deficit but by the constraints imposed by membership of the European Monetary System and international trading community. Jacques Delors, the finance minister, mouthed the fateful word 'pause' as early as 29 November 1981. At the G7 summit at Versailles in June 1982 it became clear that the other industrial countries were intent on deflation, not reflation, and Mauroy told Mitterrand that France would have to fall into line. 'If not, we will be condemned to repeat the scenarios of 1848 and 1936, when the Left was unable to govern long-term. If we do not react it will be all over in six months . . . I shall be forced to quit, like Léon Blum.'[8]

A struggle broke out between those who favoured remaining within the EMS and deflation, led by Mauroy, Delors, and Rocard, and those, led by Chevènement and Bérégovoy, who wanted to leave the EMS to pursue reflation and socialism in one country. Mitterrand avoided a decision until the municipal elections of March 1983 registered a defeat for the Socialists, and then came down in favour of the EMS and deflation. Chevènement resigned as industry minister, to be replaced by Laurent Fabius. Bérégovoy loyally followed Mitterrand's line and was put in charge of cutting the social security budget. In a dramatic U-turn, the Socialist government abandoned the 'break with capitalism' and concluded that they would have to work within the constraints of the market. Another reversal came in the summer of 1984 in the debate over education reform. Mitterrand had promised 'a great public service, unified and secular' in 1981. This was taken by militant anticlericals and the socialist party to mean cutting subsidies to private Catholic schools, and they amended the education bill of Alain Savary to this effect. At this point the Socialists felt not the constraints of international capitalism but the depth of Catholic feeling in the country. When a million people demonstrated in Paris on 24 June 1984, Mitterrand appeared on television to announce the withdrawal of the bill. Savary, who had not been consulted, resigned, and Mauroy followed him.

After the departure of Mauroy, the commitment of the Socialists to moderate policies became even more explicit. Mitterrand could have chosen Bérégovoy, Delors, or Rocard, but picked Laurent Fabius, who had headed his private office when he had been first secretary of the PS, was only 37, and had a polished, modern image calculated to win the

1986 elections. The Communists, who had been unable to put any brakes on the Socialist drift to the centre, refused to serve under him and went back to supporting the class struggle in the hope of recovering working-class votes from the PS. Presenting his policies to the National Assembly on 24 July 1984, Fabius took 'moderniser et rassembler', to modernize and unite, as his twin themes. Installing Pierre Bérégovoy at the finance ministry, he embraced the market and the mixed economy, saw the profitability of private businesses as essential, and looked to attract private capital into the public sector. He insisted on modernizing French industry, at the cost of mass lay-offs in the coal, steel, and automobile industries, but above all wanted to modernize the Socialist party to make it a good manager of the economy. Chevènement and his CERES group accused him of Americanizing the party and undertaking a French Bad Godesberg, referring to the historic conference in 1959 when the German Social Democrats abandoned Marxism. As for 'rassemblement', Fabius sought to broaden the appeal of the Socialist government by building a 'republican front', including the Centre, with a view to the 1986 elections. This tactic, however, was stamped on by Lionel Jospin, first secretary of the PS since 1981.

Cohabitation and consensus

As it became increasingly clear that the Left would lose the 1986 elections, concerns were expressed that the coexistence of a Socialist president of the Republic and a right-wing Assembly and government would throw the state back into crisis. Raymond Barre told the UDF group in parliament in September 1983 that according to de Gaulle there could be no diarchy at the summit of the state; a president faced by a hostile Assembly would either have to dissolve it or himself resign. This view was not shared by Mitterrand and neither, paradoxically, was it that of the Gaullists. Édouard Balladur, who had been secretary-general at the Élysée under Pompidou, was a close adviser of Jacques Chirac but had also dined with Mitterrand at the house of Pompidou's widow, Claude, and had already written in *Le Monde* that one day France would wake up with a president of one colour and an Assembly and government of another and had better practise 'cohabitation' rather than 'confrontation'. 'Otherwise', he reflected sagely, 'we will run the risk of transforming every political change into a crisis of regime.'[9]

In the elections of 16 May 1986 the Communists scored less than 10 per cent of the vote as their traditional supporters, incensed by the isolation and dogmatism of the party leadership, simply stayed at

home. The Socialists polled 30 per cent and lost 70 seats, leaving the
combined RPR and UDF a majority of two seats. Mitterrand had tried
to deprive them of a majority by introducing a system of proportional
representation that favoured the National Front, provoking the resignation of Michel Rocard from the government, but the 35 seats won by
the Front National were not quite enough. Mitterrand initially approached the more amenable Chaban-Delmas to see whether he could
form a government, but he did not have the support of the UDF, and
Mitterrand then observed the convention of turning to the leader of the
largest party of the majority, Jacques Chirac. For one Socialist politician this was as bad as Hindenburg summoning Hitler. The first council
of ministers was icy, as an embattled Mitterrand refused to shake the
hands of his new ministers and then clinically defined the presidential
prerogatives which he wanted respected. The new government was keen
to push through a programme of privatization, but Mitterrand insisted
that he would sign no ordinance (which required his authority) that
privatized any business that had been nationalized before 1981.

The showdown came on 14 July 1986. Chirac submitted an ordinance
on privatizations and Mitterrand refused to sign it. Chirac telephoned
Mitterrand to suggest that the presidential elections were brought
forward, then talked of resigning and provoking new parliamentary
elections. But Balladur soothed his fevered brow and advised the simple
expedient of bringing back the privatization provisions by a law, which
duly went though in August. Mitterrand had demonstrated that he was
a force to be reckoned with, but the presidency as a republican
monarchy that had been inaugurated in 1962 was clearly in difficulty.
The electorate may well have had this end in mind when it had sent a
majority of a different colour to the Assembly. Equally, though
Chirac's relations with Mitterrand were formal, they were somewhat
better than his relations with Giscard d'Estaing in 1974–6.

Two other factors, apart from the skilful mediation of Balladur,
helped to ensure cohabitation rather than confrontation. One was the
growing sense in government that there was a consensus in the country
as to which policies were acceptable and which not. 'It has not been
adequately considered', wrote Michel Rocard in 1987, 'that cohabitation would be purely and simply impossible if, beyond the division
between Left and Right that has always been caricatured, the French
people had not demonstrated a growing convergence around a few
fundamental axes of national policy.'[10] These axes were an active role
in Europe, NATO and the nuclear deterrent, which will be dealt with
in the next chapter, respect for the institutions of the Fifth Republic,

acceptance of the market, mixed economy and profitability of private business, but also acceptance of the mimimum wage and a civilized level of health and social-security provision. 'The French', another observer put it, 'have become economically liberal while remaining socially social democrats.'[11] In this respect the policies of the Chirac government differed very little from those of Laurent Fabius. Balladur resisted pressure from employers to cut the health and social-security system to a minimum safety net and rely more on private insurance. He abolished the Impôt sur les Grosses Fortunes, but later agreed that this had been a mistake. He pushed through a privatization programme, which included television, but this was interrupted by the crash of October 1987 and did not include Pechiney and Rhône-Poulenc, nationalized by the Socialists.

The second factor was the approaching presidential elections of 1988, in which both Mitterrand and Chirac intended to run. This forced them to keep a close eye on the opinion polls, which demonstrated that they were each more popular when they pulled together than when they pulled apart. This became clear during the winter of 1986–7. A bill was introduced to give greater autonomy to universities and allow them to select their own students. This went against the fundamental principles that universities were open to all students with the *baccalauréat*. University and school students protested, teachers went on strike, Trotskyists and Socialists joined the movement. On 4 December half a million people demonstrated in Paris, clashing with police, and the following day a student of Algerian origin, Malik Oussekine, died after being beaten by police. Further demonstrations took place to express outrage, joined by intellectuals such as Bernard-Henri Lévy, and in the government Balladur, who had been at Pompidou's side in 1968 and learned the art of concession, persuaded Chirac to withdraw the reform. The student protest was immediately followed by a strike movement in railways, metro, post office, gas, and electricity companies, which only underlined the comparison with 1968. Mitterrand antagonized the government by meeting a strikers' delegation on New Year's Day 1987, and Chirac made limited concessions lest the movement spread beyond the public sector. Mitterrand and Chirac emerged from the crisis only to see Raymond Barre surging ahead in the opinion polls.

The agony of Socialism

François Mitterrand approached the elections with great skill. When Chirac accused him of being a president in carpet-slippers he demonstrated that he had shown the authority of a father-figure and had acted as an arbiter on behalf of the less privileged in the 'RPR state'. Against Barre, quintessentially the man of the Centre, he presented himself as 'the president of all French people', who guaranteed a 'united France'.[12] After he defeated Chirac in the second round, he dissolved the right-wing Assembly, but stated that it would not be a good idea if one party were returned with an absolute majority. The Socialists were duly returned as the largest party, though without an overall majority, which gave the president a renewed margin of manœuvre.

Michel Rocard was made prime minister as the Socialist politician who enjoyed most credibility in non-Socialist circles. His first achievement was an 'opening' to politicians (if not parties) of the centre, bringing in ministers such as Jean-Pierre Soisson, Michel Durafour, and Lionel Stoléru, moderates who had served in the governments of Chirac and Barre, in order to ensure a working majority. He believed in modernizing socialism, abandoned Marxist rhetoric, and replaced it by a responsible managerial attitude to the economy and seeking to modernize it. Thus Pierre Bérégovoy was returned to the finance ministry to pursue 'competitive disinflation', while a moderate wealth tax was reintroduced and a Minimum Integration Income (RMI) brought in to deal with the New Poor. Rocard also believed that socialism should be less Jacobin and less statist, and allow greater autonomy for individuals and groups of citizens in civil society. To evoke a response in civil society outside the normal parameters of party politics, he brought into government a number of personalities skilled with the media, such as Brice Lalonde to be secretary of state for the environment, Bernard Kouchner, the founder of Médecins sans Frontières, as secretary of state for humanitarian action, notably in the Third World, and the Radical deputy of Marseille and chairman of its football club, Bernard Tapie.

The new presidency, which started by giving a new vigour to the Socialist party, did not continue that way. Relations between Mitterrand and Rocard had been cool since the latter had challenged the former for the first secretaryship of the party at the Congress of Metz in 1979, and deteriorated when Mitterrand decided to run for a second term as president, destroying Rocard's presidential ambitions. When Lionel Jospin was brought into the government and resigned the first secretaryship, Mitterrand tried to shoo his favourite, Laurent Fabius, into the job, not least to keep control of Rocard. But Jospin and Rocard prevented this, and secured the secretaryship for Pierre Mauroy, first in 1988, then at the Congress of Rennes in March 1990, when

faction-fighting almost destroyed the party. Mitterrand had his revenge
in May 1991, when he dismissed Rocard, but the decision did nothing
for the fortunes of the government. While keeping on most of Rocard's
ministers, Mitterrand gave the premiership to Édith Cresson, who was
regarded very much as his creature. She had little room for manœuvre,
and was hounded by the media which portrayed her as arrogant and
tactless. After disastrous results in the regional elections of March 1992,
she was in turn replaced by the long-serving Bérégovoy. Bérégovoy,
who was totally devoted to Mitterrand, was a socialist who had become
a liberal, respectful of employers as the son of a café-owner and former
gas board employee should be, and converted to the virtues of the
'franc fort' by the officials of the Finance Ministry. A left-wing Pinay,
he soon, however, began to look like a French Herbert Hoover,
presiding over economic recession and mounting unemployment. The
Socialists were punished in the elections of March 1993, when they
slumped to 18 per cent of the vote and held onto only 70 of their 282
seats. On 1 May 1993, after presiding at a ceremony at Nevers, where
he was mayor, Bérégovoy took a walk alongside a canal and shot
himself.

Driven from power a second time, the Socialist party was completely
disoriented. Michel Rocard, who had lost his seat, was elected first
secretary at the Congress of Bourges in October 1993, where he tried
to recover direction by plunging into the history of the party and
appealing to the ghosts of Jaurès and Blum. Reneging on his anti-
Communism, he also sought to build a Centre–Left alliance that would
reach from Jean-Pierre Soisson to the PCF. But the Socialist politicians
had become a caste of notables, divided into clans and riddled by
corruption, like the Opportunist republicans who had held office from
1879 to 1893. They had lost touch with their traditional supporters, the
young, the students, the working class, and the unemployed. These
deserted them in the European elections of June 1994 for Bernard Tapie
and his Énergie Radicale list. Rocard, whose list scored 14 per cent of
the vote to Tapie's 12, was forced to resign as first secretary.

Other disasters followed. Mitterrand withdrew further into the Élysée
palace to nurse his prostate cancer. Dramatic revelations in September
1994 that his right-wing past had shadowed him all the way from 1945
inflicted a massive sense of bewilderment and betrayal on the Socialist
party, and leaders like Rocard, Jospin, and Fabius, who had increasingly felt abandoned by Mitterrand, now turned their own backs on the
beleaguered president. Their one hope of salvation was that Jacques
Delors, the outgoing president of the European Commission, would

agree to run for the presidency of the Republic as a candidate of the Left. His past in the Catholic trade union movement and responsibility for the socialist U-turn of 1983 commanded support in the centre ground, and the PS, it was hoped, would rise up on his coat-tails. Unfortunately for them, Delors declined in November 1994 to run for the French presidency, arguing that even if he called new legislative elections, as Mitterrand had done in 1981, there would be no majority to support his programme, and that he had no intention of being a 'roi fainéant' or do-nothing king in the Elysée. It looked very much as if, as in 1969, no candidate of the Left would survive to the second round of the presidential elections. Lionel Jospin, the bespectacled former education minister, who was chosen as the Socialist candidate, was dubbed a 'loser' by Jack Lang, but found support solidifying around his image of modesty and probity. He confounded the opinion polls by forcing Jacques Chirac into second place on the first ballot, only to concede defeat by 47 to 53 per cent on 7 May 1995.

If it were any consolation to the Socialists, the Communists were doing even worse than they were—but only just. So touching was the loyalty of the PCF to the Soviet Union that it remained faithful to Stalinism not only after it had been abandoned by Khrushchev but after it had been dismantled by Gorbachev. Georges Marchais and the leadership of the PCF saw Gorbachev as a heretic whose reforms were responsible for the collapse of Communism first in Eastern Europe, then in the Soviet Union itself. There were, of course, critics in the party, such as Pierre Juquin and Charles Fiterman, variously called 'renovators', 'rebuilders', and 'refounders'. But under the system of democratic centralism, which had nothing democratic about it, no dissent to the line laid down by the political bureau and central committee was tolerated, and they were silenced or driven out. The party bureaucracy became increasingly isolated from the local Communist politicians and its electorate. The more it became isolated, the more it trumpeted itself as a revolutionary party of the working class and the more it became reduced to a sect. The party scored only 7 per cent of the vote in the presidential and parliamentary elections of 1988, recovering to 8 per cent in the regional elections of 1992 and 9 per cent in the parliamentary elections of 1993. It was no longer a national party but confined to the Paris suburbs, the Nord-Pas-de-Calais, Marseille, and a few rural patches in the centre, such as the Allier and Corrèze. At the 28th congress in January 1994, Georges Marchais at last resigned and the doctrine of democratic centralism was abandoned, but the party dropped back to under 7 per cent of the poll in the European elections of June 1994, and the new Communist leader, Robert Hue, managed only 9 per cent in the presidential elections of 1995.

Gaullism patrician and demagogic

The landslide of the Right in the parliamentary elections of March 1993, with 486 seats out of 577, gave them their largest majority since 1958. Jacques Chirac again asserted that the president should resign, but Mitterrand had no more intention of doing so now than in 1986. Cohabitation was no longer a leap in the dark but a cycle of French politics, and the lessons that had been learned in 1986–8 could now fruitfully be applied.

The first lesson was that Mitterrand should choose a prime minister with whom he could do business, and in Édouard Balladur he had his man. Balladur, born in Smyrna the son of a director of the Ottoman Bank, was courtly, cultivated, suave, and he seemed to have learned his diplomacy in the Orient. A practising Catholic, known as 'the canon' to his friends, his socks were bought in Rome by the wife of the French ambassador to the Holy See whereas Bérégovoy's came from Prisunic. He had an unctuous, ecclesiastical air about him and, unlike Chirac, preferred the company of duchesses to peasants. Caricatured by Plantu in *Le Monde* as a periwigged Louis XVI in a sedan chair, he was perhaps more Louis-Philippard, the last of a bourgeois dynasty, shaped by the *grands corps* and the boardroom, a master of the imperfect subjunctive while his interior minister, Charles Pasqua, mouthed the language of the people.

Balladur had oiled the wheels of the first cohabitation and he ensured that they ran smoothly during the second. The fact that the ageing and sick Mitterrand had no plans to run for the presidency in 1995, whereas Balladur was a leading contender, ensured that the rivalry that had divided Mitterrand and Chirac was not repeated. Balladur's rivals for the presidency were all within the majority. Philippe Séguin, the speaker of the National Assembly and author of a book on Napoleon III, took the Bonapartist view that Balladur was protecting the privileged rather than the common people, and accused him in June 1993 of a 'social Munich'. Jacques Chirac, who declined to serve in the Balladur government, fell back on his leadership of the RPR and, learning from Séguin, recast his appeal in the light of the original principles of Gaullism or Bonapartism with a view to challenging for the presidency.

Until the presidential campaign gathered momentum, however, the main problem for Balladur was that the parliamentary majority was so powerful that his government was under pressure from the Right to drive through policies that violated the national consensus. Because left-wing opposition in parliament was so insignificant, the only resistance to this pressure came from within the majority, from the Conseil Constitutionnel, which was chaired by Mitterrand's former minister of justice, Robert Badinter, from periodic election setbacks, and from the streets. Within the majority, the centrist leaders Simone Veil, now social affairs minister, and Pierre Méhaignerie, the minister of justice, objected to Pasqua's plans to strengthen police powers to check the identity of foreigners, on suspicion that they might be illegal immigrants, but they were forced by the weight of the parliamentary majority to toe the line. On the other hand, Pasqua's immigration bill was modified by the Conseil Constitutionnel in August 1993 on the grounds that it infringed individual liberties, forcing the minister of the interior to seek a revision of the constitution on the question of the right of asylum. The Conseil Constitutionnel also intervened in the matter of the revision of the Loi Falloux.[13] The Loi Falloux of 1850 was itself a charter for private education, but it limited the extent to which local authorities could finance private schools. A reform of the law, to open the way to additional public funds for private schools, was demanded by the Right and rushed through parliament in December 1993. This provoked a strike of teachers in the public sector, and was criticized both by François Mitterrand and by Mgr. Decourtray, cardinal-archbishop of Lyon, who feared that it would reopen the 'school war'. On 13 January 1994 the reform was censured by the Conseil Constitutionnel for violating the principle of equality between citizens, and a mass demonstration of 900,000 in Paris on 16 January turned into a victory parade.

In one other sphere the national consensus was maintained not by the government or the Assembly but in the streets and in the polling booth. The Balladur government was faced on the one hand by rising unemployment, particularly of young people, and on the other by pressure from employers to relax the constraints of the SMIC or minimum wage. Its answer to this was the Professional Integration Contract (CIP) or 'SMIC-Jeunes', which permitted employers to pay young people 80 per cent of the minimum wage. This project was opposed by university and school students, teachers, and all the unions, both Communist and non-Communist, which co-operated for the first time since 1980. Demonstrations swept Paris and the provinces on 17 and 22 March 1994, often erupting into violence and once more brandishing the spectre of 1968. The riots were compounded by a poor showing of the majority in the cantonal elections of 20/27 March 1994, in which the

Socialists bounced back to 23 per cent and the Communists to 11 per cent. Though this result turned out to be a flash in the pan, Balladur took it as a salutary warning and shelved the CIP on 28 June. Édouard Balladur was not Margaret Thatcher, nor did he make every social conflict a test of his machismo. He saw himself as the heir of Georges Pompidou, to whom he paid tribute in the Cantal on the twentieth anniversary of his death in April 1994, reflecting:

The authority and even the prestige of the state have nothing to gain from defending decisions which are not understood by the population, come hell or high water. Negotiation, dialogue, reciprocal good faith and the sense of the national interest must make it possible to find solutions to difficulties, even at the cost of additional delays.[14]

At the beginning of the presidential campaign of 1995 it seemed as though the benign Balladur would move effortlessly from Matignon to the Élysée. He resurrected the Gaullist idea of 'rassemblement', winning support from the UDF, and broadcast the slogan 'It's safer with Balladur'. However, while he would have wished to concentrate on the Socialist opposition, he was forced to deal first with the challenge of Jacques Chirac who, emerging from his tent for a third bid for the presidency, threatened to split the Gaullist movement.

As the Chirac offensive gained momentum, so Balladur's safe pair of hands looked unequal to the task of dealing with the besetting problems of unemployment, poverty, and social dislocation. Chirac attacked Balladur's defence of the status quo and somnolent style of government, and countered it with the Bonapartist discourse that had been revived within the RPR by Philippe Séguin. Thus he attacked the élites and technocrats who ruled France and were increasingly out of touch with the people, offering himself as a strong leader who would make frequent use of the referendum. He spoke of a 'republican pact' that made the Republic the 'thing of all' and would bring back the dispossessed within its ambit by means of investment, job-creation and wage-rises. To make his point he left the plush surroundings of the Paris Hôtel de Ville and tirelessly toured the regions of France to press the flesh and drink beer, meeting peasants, workers, and the unemployed. At the same time, however, he mouthed liberal ideas presented to him by Alain Madelin, the UDF business minister who in the 1960s had been a member of the extreme-Right Occident movement. This bound Chirac to a policy, more or less in contradiction with his Bonapartism, of reducing the government deficit, cutting taxes, and lightening the burden of social insurance on businesses. It was designed to appeal not to the poor or unemployed but to the private sector of small and medium businesses, traditional supporters of the Right.

Balladur was obliged to respond by coming off his prime ministerial pedestal and descending into the electoral arena. He accused Chirac of demagogy, of threatening the strategy of the strong franc, and of being a socialist mole within the government majority. But Chirac's demagogy appealed disproportionately to young men, Balladur's patrician airs correspondingly to old women. Thus Balladur received only 18 per cent in the first ballot, against Chirac's 21, and was eliminated from the race. The astuteness of Chirac's campaign now became clear. Though a man of the Right, he correctly assessed that the battle would be fought on the terrain of the Left, on the issues of unemployment and poverty that two terms of a Socialist president had been unable to alleviate. Though the Gaullists and the UDF now united behind him, he gauged that the necessary consensus had to be built not among the political class but, appealing over the heads of the political class, among the people itself. His populism was 'social' without being socialist, attacked bureaucrats and experts, drew a classic distinction between productive and speculative capital, and promised a tough line on law and order and immigration. Though incoherent at the level of policy, it offered at least the rhetoric of action to rescue France from division and disillusionment, and finally delivered him the coveted prize of the presidency.

The challenge of the extremes

The experience of pluralism with the alternation and cohabitation of Right and Left in power, the growth of a consensus around key issues, the displacement of totalizing ideologies by a pragmatic liberalism, and the sober management of the affairs of the mixed economy were all of positive benefit to the Republic, which now no longer seemed to be in danger. There were, however, other ways of looking at the same phenomena. Pluralism meant the swapping of power between professional politicians whose sole concern was to cling to office. The consensus of the Republic of the Centre meant the exclusion of the extremes, who were deprived of access to the media and portrayed as lunatic or dangerous. Obsession with the narrow economic and social concerns of managing the mixed economy sidelined other issues, such as immigration, national identity, or the environment.

After 1981 two important movements emerged to challenge this cosy consensus: the National Front and the Greens. At first sight it may be doubted whether they had anything in common. But they represented two attempts radically to challenge the established consensus, and to set out a new agenda for politics which went far beyond the eternal issues of the economy and welfare system. They were both conservative, backward-looking, and opposed to technological change. They both had a totalizing vision of the world and were committed to the ideas of purity and regeneration, whether of the nation or nature. They attacked all politicians as equally guilty, and sought to define new forms of political action. In their different ways they came up against two problems. First, the extent to which it was possible to engage with political activity without becoming forced to play the political game according to the rules laid down by the politicians in power and becoming corrupted by the existing political system. Second, the extent to which they could force their agenda into the mainstream of political debate and, if they managed to do that, to ensure that the political capital benefited them and not established politicians who sought to absorb the new issues into the conventional political discourse.

The National Front

The eruption of the National Front onto the political scene after 1981 signalled the transformation of the extreme Right from a battery of small movements involved in direct action into a party that was a genuine challenger for power, bringing together extremists and those who had abandoned traditional parties of the Right. Founded in 1972, it made its breakthrough after the triumph of the Left and the corresponding humiliation of the traditional Right in 1981. It won 11 per cent of the vote and 10 members in the European elections of 1984, 10 per cent and 35 deputies in the parliamentary elections of 1986, 14 per cent in the presidential election of 1988, and 10 per cent (but only one deputy under the first-past-the-post system) in the succeeding parliamentary elections. It won nearly 12 per cent in the European elections of 1989, and 14 per cent in the regional elections of 1992, falling back to 12 per cent in the parliamentary elections of 1993 and 10 per cent in the European elections of 1994, but returning to 15 per cent in the presidential elections of 1995.

The National Front was able to capitalize on the failings of the other parties, but its electorate had the weaknesses as well as the strengths of a protest vote. Sociologically, it fluctuated considerably. In 1984 it received support from the rich bourgeoisie, who deserted the classic Right after its failure in 1981, and scored 17 per cent in the sixteenth *arrondissement* of Paris and 18 per cent in Neuilly-sur-Seine. In 1986 its voters were more working-class or unemployed, living cheek-by-jowl with immigrant populations. In 1988 Le Pen managed to combine working-class and petty bourgeois voters, and the profile of his supporters, compared with those of other candidates, was more male, less educated, more inclined to live in cities of over 100,000, and less religious than others on the Right. The electorate of the National Front was apolitical, floating, and not altogether determined to see the party in power. In 1986 a third of its supporters had voted for Mitterrand in 1981 and another third had abstained. Whereas a quarter of all voters polled after the 1988 presidential elections had voted in the past for candidates of three or more parties, this was the case of 59 per cent of Le Pen's voters; and whereas over 80 per cent of supporters of Barre, Chirac, and Mitterrand wanted 'from the bottom of their heart' to see their candidate president, this was the case of only 28 per cent of Le Pen supporters.

The novelty of the National Front agenda was not so much its obsession with immigration, crime, and taxation as its refusal to admit the possibility that immigrants might be assimilated into French society. It held that immigrants were responsible for all France's problems, from the housing crisis and unemployment to rising crime, AIDS, and the undermining of the French nation, and that immigrants should therefore be repatriated. Le Pen himself always denied that he was racist, asserting only that he had a natural preference for his family, colleagues, commune, province, and nation to foreigners, and that the interests of the French must come first. He denied also that he incited racial hatred, arguing that his enemies were the politicians of the established political parties, who failed to deal with these crucial issues, and the media, which waged a campaign of silence and vilification against him. He was as critical of the classic Right as he was of the Left, calling it (taking the slogan from Guy Mollet) 'the stupidest in the world', attacking it both for letting the Left win in 1981 and 1988 and for cohabiting with it in 1986 and 1993.

Le Pen and the National Front were populists who appealed to the people against the politicians. His straight-talking views evoked sympathy not only from those who voted for him but those who did not but confessed to sharing his ideas, a proportion that rose to 32 per cent in an opinion poll of October 1991. Bruno Mégret, a leading National Front politician, claimed the previous month that 'all particles of political life are being influenced by the magnetic field created by the FN . . . henceforth the FN is dominating political life.'[15] But from September 1983, when Jean-Pierre Stirbois became deputy-mayor of

Dreux after the National Front were allowed onto a joint list with the RPR and UDF, the National Front had needed the support of the traditional Right to gain seats and local office. Embarrassed by some of these alliances, many on the traditional Right were not at all embarrassed to steal the political clothes of the National Front in order to win back supporters they believed had been lost to the Front. With Pasqua at the ministry of the interior in both 1986 and 1993, the Right took up Le Pen's ideas on immigration control and the redefinition of French nationality. Le Pen replied by attacking the pusillanimity of the Balladur government, and denouncing it as worse than that of Bérégovoy. But so unconvincing were these criticisms that in January 1994 a third of National Front supporters thought that the party should support the government, only 11 per cent wanted it to attack, and 47 per cent were indifferent. The same poll showed that the proportion who shared his ideas had dropped to only 23 per cent, while 73 per cent—compared to 38 per cent in October 1983—saw them as a threat to democracy.[16] Despite the 15 per cent poll of Le Pen in the presidential elections of 1995, the same constraints were evident. On the one hand, despite Le Pen's dramatic refusal to support Jacques Chirac in the second round, half his supporters were happy enough with Chirac's views on immigration to do so. On the other, the death of a Moroccan, thrown into the Seine by skinheads who then melted into the National Front parade of 1 May, provoked a wave of anti-racism among those determined to contain the National Front.

The Greens

For a long time the Greens were not a political party, and when they did form one, they claimed it was not like other political parties. The movement started among the *gauchistes* of 1968, including Brice Lalonde, who formed Friends of the Earth in 1970. Its prime target was the government's nuclear programme, both civil and military, and it practised direct action, culminating in the demonstration of 60,000–80,000 on 31 July 1977 to stop the building of a nuclear super-reactor at Creys-Malville, near Grenoble. The movement attracted interest from and overlapped with the anarchists, the CFDT, and Michel Rocard's PSU, but the Greens criticized the labour movement as productivist and polluting and feared political hijacking by political parties, while anarchists believed that the movement was drifting into politics and away from the libertarian revolution. At first the movement organized politically only sporadically, to run René Dumont in the presidential elections of 1974, and Brice Lalonde in those of 1981 or to fight the European elections of 1979. When it did start to organize as a party after 1979, in order to retain and structure its support, the Friends of the Earth objected. The Greens were constituted as a united party in January 1984 but almost immediately expelled Brice Lalonde, who ran his own list in the European elections that year, to prevent the Greens obtaining the threshold of 5 per cent they required to win seats.

Even as a political party, the Greens asserted that they were different. Their structure was loose and extremely democratic, with spokespeople rather than leaders. Their agenda was totally new, postulating a contract between man and nature equivalent to the social contract between man and man. They defined themselves as neither on the Right nor on the Left but 'forward'. Their membership was politically virgin—only 15 per cent had belonged to other political parties or movements in 1988—and had a rapid turnover. The profile of their electorate was very distinct: young and educated, overwhelmingly students, teachers, and public-sector workers, strong in certain regions such as Alsace, Lorraine, Franche-Comté, Brittany, Normandy, and the Île de France. Despite the refusal of the leadership to choose between Right and Left, three-quarters of those who voted Green in the European elections of 1989 had voted for Mitterrand in 1988, and the party drew heavily on the 'DDS' (*Déçus du Socialisme*, or Disillusioned with Socialism).

The persistent argument within the Green movement was how to relate to other political parties, especially to the Socialists. One tendency, represented by the serious, Alsatian Catholic Antoine Waechter, was that the Green party must never do any deals but remain completely autonomous in order to defend its identity and integrity. Another, represented by militants of teachers' unions such as Yves Cochet, argued that the Greens must be ready to ally with regionalists, feminists, anti-racists, champions of the Third World, even with Socialists, in order to win seats. In this way they won over 10 per cent of the vote and 9 seats in the European elections of 1989. The argument developed into one concerning how they should relate to power when Brice Lalonde was appointed Michel Rocard's secretary of state for the environment in 1988. Lalonde was exasperated by the purism of Waechter's Greens and believed in 'Greening' the existing political system. He was very active internationally, negotiating agreements on the ozone layer, greenhouse gases, mining in the Antarctic, and the ivory trade, but he was not in a position to criticize the government's nuclear tests in the Pacific or the pursuit of war in the Gulf. The Socialists were happy to steal individual policies from the ecologists, but not to let them interfere with the overall thrust of their strategy.

In May 1990 Lalonde launched his own movement, Génération Écologie. Denouncing Waechter as an Alsatian hick and the Greens as a small group of Cathars or a neo-National Front, he sought to win converts from the existing political class and appealed to 'realistic ecologists, reforming centrists and modern socialists.'[17] At first it seemed that Génération Écologie would serve to undermine the Greens and widen the faltering presidential majority. But after his eviction from the government by Bérégovoy in 1992 Lalonde sought to attract the DDS, and even made an electoral pact with Waechter's Greens for the parliamentary elections of 1993. Fully expecting to win 17 per cent of the vote, they crashed with 4 per cent for the Greens and 3 per cent for Génération Écologie.

Lalonde then showed his true opportunistic colours by accepting a foreign trade mission from Édouard Balladur. He argued that the Balladur government was further to the Left than that of Bérégovoy while denouncing *'gauchiste* cancers' in his own party and purging dissidents such as Harlem Désir. Lalonde's strategy of 'Left or Right' was far from persuading all his supporters. In the Green party, meanwhile, those such as Yves Cochet and Dominique Voynet, deputy for the Jura, who favoured allying fairly and squarely with the Socialists, not least after running the Nord-Pas-de-Calais region together since 1992, defeated the 'Khmers verts' at the Lille congress of November 1993. Waechter, still holding to his 'neither Right nor Left' line, was ousted as leader and replaced by Dominique Voynet. He went on to found his own Independent Ecologist Movement in 1994. But those who imagined that political deals with Rocard's Socialists would serve the Greens better than ideological purity were mistaken. In the European elections of 1994 the Greens slumped to 3 per cent, Génération Écologie to 2 per cent. In the presidential elections of 1995, despite the unpopularity of the mainstream candidates, Dominique Voynet still only attracted 3 per cent of the vote.

Political alienation and the new politics

If the greatest threat to the political class in the 1980s was the rise of the National Front and the Greens, that of the 1990s, once these challengers had peaked and begun to decline, was the growing disdain in which it was held by the electorate. This political alienation or crisis of representation was illustrated first by falling rates of political participation among citizens, especially among young people, secondly by the growing opinion that all politicians were incompetent, divided

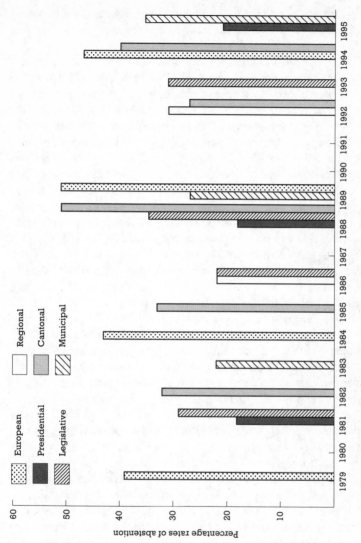

FIG. 9. Rates of Abstention of the French Electorate, 1979–1995.

by faction, and corrupt, and thirdly by the rise of a new kind of politics that had more to do with show business and the mass media than with traditional parties and politicians.

The proportion of those actively involved in political parties and movements was never large, but figures for 1988 suggested that only a million people, or 2–3 per cent of the population, belonged either to a political party or to a movement such as ecologists or feminists. The membership of trade unions, which had been 23 per cent of the working population in 1975, fell to 12 per cent in 1990. The proportion of those reading a daily newspaper fell from 55 per cent in 1973 to 43 per cent in 1988, compensated by a rise in the readership of magazines. Even among those whose sole political act was to go into the ballot box on election day, the rate of abstention in elections was rising. Presidential elections, with the emphasis on personalities, remained popular, and the abstention rate was only 19 per cent in 1981 and 1988 rising to 21 per cent in 1995. Municipal elections, involving local notables, and direct regional elections, which articulated some sense of provincial identity and were first held in 1986, were more popular than cantonal elections, to elect members of the departmental *conseil général*, in which the abstention rate reached 51 per cent in 1988. Overall, however, in all kinds of elections, the abstention rate was rising. In elections to the National Assembly, the abstention rate was 20 per cent in 1968, 19 per cent in 1973, and 17 per cent in 1978, but it was 29 per cent in 1981, 22 per cent in 1986, 34 per cent in 1988, and 31 per cent in 1993. In regional elections the abstention rate was 22 per cent in 1986 and 31 per cent in 1992, while in European elections it was 39 per cent in 1979, 43 per cent in 1984, 51 per cent in 1989, and 47 per cent in 1994.

Between 1956 and 1988 rates of abstention grew fastest in northern France, in heavily urbanized departments, and in areas such as Lorraine and the Loire, suffering economic crisis. They tended to reflect both the concentration of young people and rates of unemployment, which in any case affected young people more. What might be called passive abstention expressed a total lack of interest in politics; active abstention, on the other hand, expressed disillusionment with the political process and was often a form of protest. France had suffered twenty years of indifferent economic performance, and neither governments of the Left nor governments of the Right seemed able or willing to do anything about it. Whereas until about 1969 the regime was seen as crucial either to protect individual rights or to secure order, now that the Republic was safe from either revolution or reaction, citizens saw the government rather as a threat to individual rights and happiness, and sought to limit interference by the state and broaden the sphere of a civil society seen to be independent of the state.

Politicians were seen not only as meddlesome, factious, and incompetent; they were regarded as corrupt. In November 1990 a Sofrès poll indicated that 55 per cent of French people thought their politicians were corrupt. To some extent this was facilitated by the political system. In the first place French politicians tended to hold both national and local office, being deputies or senators, mayors of large cities, chairmen of their *conseil général* or regional councils, and members of the European parliament. Influence in Paris or Brussels was crucial to ensure funds and contracts to develop their locality while, conversely, a local power base or fief was necessary to ensure election to parliament. This *cumul* or multiple office-holding grew like a cancer: whereas 36 per cent of deputies practised it in 1946 and 42 per cent in 1956, the figure rose to 70 per cent in 1970 and to 96 per cent in 1988. Second, the massive development of the economic infrastructure after 1945 and the shift towards administrative decentralization after 1981 put immense power and resources at the disposal of local politicians. Mayors, deputies, and chairmen of *conseils généraux* and regional councils also sat on semi-public companies or quangos which awarded valuable contracts, and operated outside the knowledge and supervision of democratically elected assemblies. Third, however, the need for election expenses was such that the financing of politicians and parties by businesses was extremely welcome. Here was the main source of political corruption: that contracts for local and regional development were given in the expectation of a contribution to election expenses or blatantly as a reward for them. Even more glaring an abuse was that money intended to support party finances found its way into the private accounts of politicians.

Accusations of corruption were made in the first instance against local politicians. Michel Noir, elected mayor of Lyon in 1989 and until then one of the hopeful young 'renovators' of the RPR, François Léotard, deputy mayor of Fréjus and leader of the Republican Party, and Maurice Arreckx, also in the Republican Party, mayor of Toulon and 'godfather of the Var', heading its *conseil général* between 1985 and 1994, were among those pursued by the courts. Most notorious was Jacques Médecin of the Independents, who between 1966 and 1990 ran Nice like a Mafia city and siphoned funds from a straw company, Nice Opera, supposed to scout for musical talent in the United States, into private bank accounts in California and Panama. Exposed in 1990, he fled to Uruguay in 1990, sold T-shirts at Punta del Este before buying a hacienda, and was arrested in 1993.

Corrupt politicians were not all local and not all of the Right. Socialists were no less guilty of corruption, the highest level included. Links between government and business, particularly in respect of takeovers and privatizations, were conduits for corruption. After vast profits were made by the takeover of American Can by Pechiney in 1988, accusations of insider dealing were made against Patrice-Roger Pelat, an old friend of François Mitterrand, Max Théret, a press magnate and important source of Socialist party funds, Samir Traboulsi, a Lebanese businessman, and Alain Boublil, the head of Pierre Bérégovoy's private office. Bérégovoy was later said to have received an interest-free loan from Pelat to buy a Paris flat. Pelat died in 1989 as the scandal raged, Théret, Traboulsi, and Boublil all received prison sentences in 1993, and Bérégovoy's suicide was as much connected with rumours of corruption as with the Socialist defeat, of which it was a major cause. The hands of the incoming Balladur government, however, were no cleaner, and in 1994 three ministers were forced to resign as charges of corruption were levelled at them. Alain Carignon, RPR minister of communications, was accused of privatizing the Lyon Water Company as a reward for its funding of his successful bid to become mayor of Grenoble in 1989, and sent for trial in 1995. Michel Roussin, RPR minister of co-operation, was said to have accepted bribes from a large development company in the Paris region while he had been Chirac's chief adviser at the Paris Hôtel de Ville. Most importantly, Gérard Longuet, the industry minister, was forced to resign after accusations that as treasurer of the Republican Party he had channelled funds from the property company Cogedim to pay for building his villa in Saint-Tropez.

Party politicians dealt in a spasmodic and perfunctory way with the problem of corruption. Just before the elections in 1986 the Socialists passed a law limiting the number of elective offices that could be held together to two, not counting being mayor of a town of fewer than 20,000 inhabitants. Similarly, just before the elections of 1993 they passed a law regulating the financing of parties by businesses, in an attempt to eliminate favouritism in the granting of public contracts. Late in 1994 Philippe Séguin, as president of the National Assembly, put himself at the head of the public outcry against corruption and promoted a bill to outlaw favouritism in the granting of public contracts and end the financing of political parties by businesses. But while he had the support of Socialists and Communists, the Right in general opposed him, and even Séguin admitted that there was no possibility of getting a bill abolishing pluralism through parliament.

One response to corrupt and incompetent politicians was abstention. Another was to vote for a party like the National Front, which used the issue of corruption in order to attack all the mainstream political parties together. Finally, there was the evolution of a new form of politics. This was characterized by enthusiasm for a charismatic figure, often of the world of show business or rock music, dealing with alternative issues such as racism, unemployment, or the Third World, hyped by the mass media and finding a direct response with the young outside the channel of the established political parties. Three examples may suffice. The first was that of the comedian and clown, Coluche, who tried to run in the presidential election of 1981. Supported by intellectuals such as Pierre Bourdieu, Gilles Deleuze and Félix Guattari, the authors of *L'Anti-Œdipe*, and by film stars like Belmondo and Dépardieu, promoted by the satirical magazine *Charlie-Hebdo*, he appealed to 'the idle, the unkempt, drug-addicts, alcoholics, gays, women, parasites, the young, the old, artists, jailbirds, prostitutes, apprentices, blacks, pedestrians, Arabs, French people, hippies, madmen, transvestites, ex-Communists, hardened abstentionists, all those who don't count for the politicians.'[18] Unfortunately, he was squeezed out by the political class, which denied him the 500 signatures of elected representatives he needed for his candidature, and by the television, which refused to interview him, and he was forced to withdraw from the race.

Five years later Coluche was back, acting as a patron for SOS-Racisme. SOS-Racisme had the allure of a spontaneous youth movement of solidarity with the Beurs, but in fact it was carefully packaged and promoted and linked the worlds of politics, rock, and the media. The power behind the throne was the young socialist militant, Julien Dray, whose contacts included Jacques Attali and Jean-Louis Bianco at the Élysée. Harlem Désir was the media-friendly frontman, Coluche and Bernard-Henri Lévy had links with the media. The yellow-hand badge and 'Touche pas à mon pote' slogan was launched at a press conference on 22 November 1984, and a rock concert was organized on the place de la Concorde on 15 June 1985. In the end, however, the Beurs themselves drifted away as SOS-Racisme increasingly became a springboard for party politics. Julien Dray was elected a Socialist deputy for Essonne in 1988, while Harlem Désir ran unsuccessfully as a Génération Écologie candidate in the parliamentary elections of 1993.

The final example is that of Bernard Tapie. Tapie was an electrician and former pop star who became a wealthy tycoon by setting up a holding company in 1979 which, among other concessions, owned

Wonder batteries and Wrangler jeans in France. He sustained a
flamboyant lifestyle, collecting eighteenth-century art and furniture,
mooring his yacht in Marseille harbour, and buying Olympique Marseille football team in 1986. Elected deputy of the Bouches-du-Rhône
in 1988, he served briefly in the Socialist government. Both he and his
football team were corrupt, but his corruption was seen as somehow
thumbing his nose at the establishment. The fact that the courts
confiscated his yacht and furniture for tax evasion and the Crédit
Lyonnais pursued him for debts only served to confirm his Robin Hood
status with his have-not supporters. He was the antithesis of the
'governmental' Socialist Michel Rocard: in revolt against the taxman,
the big banks, the judicial system, a tribune of the people emerging
from the ruins of the classic Left. In the European elections of 1994,
even while he was being pursued through the courts, his Énergie
Radicale list won 12 per cent of the vote and finished the political career
of Michel Rocard. The withdrawal of Jacques Delors from the race for
the presidency of the Republic in December 1994 made Tapie look like
a possible candidate of the Left, until he was declared bankupt and
ineligible to stand for elective office. Tapie was politics as soap, now up,
now down, but always fighting back and always in the public limelight.

8

France in Search of a World Role

Valéry Giscard d'Estaing had the honesty to confess during his presidency that France was now a medium-sized power. This did not mean that she no longer had pretensions to grandeur. She fought hard to retain her permanent seat on the Security Council of the United Nations, struggled to ensure her hegemony in Europe, and proved both stubborn in relinquishing the last vestiges of her empire and remarkably inventive in perpetuating different forms of neo-colonial power. Over time, however, many of these ambitions bore bitter fruit. When Soviet power collapsed, France found it increasingly difficult to stand up to the United States. The unification of Germany threatened her leading role in Europe and cooled the European fervour of many French people. In Africa the French lost their ability to arbitrate effectively, while within Francophonie they gradually lost the initiative to a new rival, Canada.

Standing up to America

When de Gaulle returned to power in 1958, his ambivalent attitude to the United States was little changed. He both respected and resented the Americans for having liberated France, was both grateful that the United States provided an umbrella of protection for the free world and hated the fact that the United States alone had its finger on the button of the West's nuclear arsenal and put her own strategic interests before that of her partners in NATO. De Gaulle was intent on forcing his way into the club of superpowers. In September 1958 he suggested to President Eisenhower that decisions on the use of nuclear force by NATO should be taken not by the United States alone but by a three-power directorate of the United States, Great Britain, and France. In March 1959, after Eisenhower had rejected this proposal, de Gaulle removed the French Mediterranean fleet from the control of NATO, arguing that NATO's sphere did not extend to the southern Mediterranean and therefore could not protect France's interests in

North Africa. At the same time France pressed ahead with developing its own nuclear deterrent, over which de Gaulle as president of the French Republic would have total control. This would both underpin French sovereignty and ensure the safety of France's place as a permanent member of the Security Council of the United Nations. The French exploded their first atomic bomb in the Sahara desert in February 1960. The Americans tried to starve the French nuclear programme by depriving it of information and technology, and in December 1962 President Kennedy agreed to sell Polaris missiles to Great Britain. In response to the perceived Anglo-Saxon menace, de Gaulle not only vetoed the British application to join the Common Market but refused to sign the nuclear test ban treaty endorsed by the United States, Great Britain, and the USSR in July 1963.

In February 1966 de Gaulle announced at a press conference that France was withdrawing from the integrated command structure of NATO, and asked the Americans to remove their thirty bases and 26,000 troops from French soil. As de Gaulle put it that October,

there is no longer any actual or possible subordination of our forces to a foreign power. In six months there will no Allied command, unit, base or army on our soil. We will restore their wholly national character to our army, navy, air force in matters of command, operations and training.[1]

Fears were expressed in parliament that the United States would no longer defend Europe, and that without American troops another defeat like 1940 might not be avoided. The situation, however, was not entirely clear-cut. France remained in the NATO alliance, which committed its partners to fight in the event of an act of unprovoked aggression committed against one of them, and de Gaulle fully supported Kennedy during the Cuban missile crisis. Richard Nixon, who idolized the General, visited him in Paris in January 1969, while Michel Debré, as defence minister, took part in the celebrations to mark the twentieth anniversary of NATO in Washington, just before de Gaulle fell from power.

France not only remained loyal to NATO, she also became increasingly an economic vassal of the United States. American investment in France shot up in 1961 and peaked in 1962–5, saturating key industries such as chemicals, engineering, electrical goods, farm machinery, and food-processing. In January 1963 Chrysler bought a controlling interest in Simca, France's third largest car manufacturer. Later that year General Electric made a bid to buy up the French computer firm Bull and, despite attempts by the French government to block the move, succeeded. Even the French language was being colonized by English, protested the Sorbonne professor René Étiemble in *Parlez-vous franglais?* (1964). But in another highly popular work, *The American Challenge*, published in 1967, Jean-Jacques Servan-Schreiber argued that the answer was not to stop American investment but to beat the Americans at their own game by copying their skills in science, technology, and management and building an effective partnership between business, the universities, and government.

Meanwhile the response of de Gaulle was to engage in nineteenth-century diplomacy and renew the Franco-Russian alliance, this time against America rather than Germany. In June 1966 he visited the Soviet Union, recalling his previous trip in 1944 and claiming that no fundamental interests had ever separated France and Russia, even at the time of *War and Peace* and Sebastopol. He spoke of European co-operation from the Atlantic to the Urals, which rather bemused his Soviet hosts, who had no idea their country was so small. De Gaulle also visited other Eastern bloc countries: Poland in September 1967 and Romania in May 1968, before the Soviet invasion of Czechoslovakia in August 1968 demonstrated that the Cold War had not ended and the blocs were as solid as ever.

Parallel to this, de Gaulle undertook a world tour in order to denounce American imperialism. He pointedly recognized America's bugbear, the People's Republic of China, in January 1964. He visited the American preserve of Mexico in March 1964 and, stopping off in Guadeloupe on the way home, announced that the French were still a great nation. He went to Canada to demonstrate that she was not a client state of the United States, and in Montreal in July 1967 famously declared 'Long live free Quebec!' Though the Americans had only recently taken over from the French in Vietnam, de Gaulle addressed a crowd of 100,000 in a sports stadium at Phnom Penh on 1 September 1966 posing as the man who had given independence to the Algerians and attacking the escalation of the war in Vietnam by Lyndon Johnson. Lastly, de Gaulle denounced America's ally, Israel, in the Six-Day War of June 1967, supported a call for her withdrawal from the occupied territories in the UN, and imposed an embargo on arms supplies.

In the Middle East the French saw it as their interest to support the Arabs rather than the Israelis. President Pompidou continued the arms embargo against Israel while concluding a major contract to sell arms to the new Libyan regime of Colonel Gadafy in November 1969. In the light of this, the presidential visit to the United States in February 1970 was a disaster. M. and Mme Pompidou were heckled and jostled by demonstrators when they turned up for dinner with Mayor Daley in Chicago; she took the next plane home, and he never set foot in the States again, subsequently meeting Nixon outside France, in the Azores or Iceland. The outbreak of the Yom Kippur war between Egypt, Syria, and Israel in October 1973 did nothing to improve Franco-American relations. The United States and most of the West supported Israel, and OPEC raised its oil prices to force them to relax that support. The French, who relied on oil for three-quarters of their energy supplies, three-quarters of that oil coming from the Middle East, could not afford to alienate the Arab states. On the contrary, they could only meet their oil bill by selling what the Arab states most wanted, namely arms. So Foreign Minister Jobert visited Syria, Iraq, Libya, and even American's client, Saudi Arabia, in January 1974, selling arms for oil. As a result, 55 per cent of French arms exports in 1974 went to the Middle East.

While the relations of Pompidou and his foreign minister with the United States were execrable, their relations with the Soviet Union were excellent. Pompidou visited the Soviet Union in 1970, 1973, and 1974, while Brezhnev came to France in 1971 and 1973. Giscard d'Estaing, by contrast, was widely seen to be an 'Atlanticist'. He removed Jobert from the Quai d'Orsay and was fulsome in his praise of the Atlantic alliance. Behind the scenes the Americans began to help the French with nuclear information and technology, in recognition of the contribution of the French deterrent to NATO as a whole and on the understanding that, even if France did not return to the integrated NATO command, she would co-operate over the use of her deterrent in the event of war.

This is not to say that France jumped into the pocket of the United States. The Middle East remained a stumbling block, not least as France sought to realize the benefits of exporting her nuclear technology. In November 1975 France agreed to equip Iraq with a nuclear reactor that was supposed to be for peaceful purposes alone. Foreign Minister Sauvagnargues met the PLO leader Yasser Arafat in Beirut in October 1974 and the PLO were allowed to open an office in Paris the following year, their claim to statehood supported by the French government. Giscard attempted to create a privileged relationship with Anwar Sadat, who visited Paris in January 1975 and received Giscard in Egypt the following December. Giscard dreamed of a conference arbitrating French, African, and Arab affairs, and resented President Carter's stealing the limelight by brokering an agreement between Sadat and Begin at Camp David in March 1979. While the Americans propped up the crumbling regime of the Shah of Iran, the French welcomed Ayatollah Khomeini to France in October 1978, sent him back to Teheran on an Air France jet in February 1979, and did little to protest when American hostages were taken prisoner the following November.

Giscard met Carter only once, on the Normandy beaches in January 1977, when Carter enquired about his much-publicized informal lunches with ordinary French families. He was more energetic in his cultivation of Brezhnev, whom he received at Rambouillet in December 1974, and whom he visited in Moscow in October 1975 and April 1979. Giscard prided himself on the pragmatic nature of his relations with the Soviet Union, and privately criticized Carter for letting questions of ideology and human rights interfere. Following the Soviet invasion of Afghanistan in December 1979, Giscard resisted demands from the United States for an embargo and refused to order a boycott of the 1980 Moscow Olympics by French athletes. For him, the negotiation of a natural-gas pipeline from Siberia to the West was far more important. He went as far as to meet Brezhnev once again, in Warsaw, on 19 May 1980, prompting jibes from his rival for the presidency, François Mitterrand, that he was no more than Brezhnev's telegraph boy.

Going along with America

It might be imagined that, in the area of foreign and defence policies, the coming of the Left to power in May 1981 would represent a fundamental shift. Mitterrand had attacked de Gaulle's independent nuclear deterrent when he ran against him in 1965, and the common programme of Socialists and Communists renounced the use of nuclear force by France and called for universal nuclear disarmament. Similarly, the Left was traditionally no friend of the United States, seeing it as the source of economic, military, and cultural imperialism.

At first there were signs of an anti-American foreign policy. Presenting himself as the champion of liberation and non-alignment, Mitterrand visited Mexico in October 1981, Algeria two months later, and India in November 1982. A joint Franco-Mexican declaration of August 1981 proclaimed solidarity with the rebels of El Salvador fighting the US-backed junta of Napoleon Duarte. The following year, the French signed an arms deal with Nicaragua to help beat off the Contras and offered to help clear CIA mines in Nicaraguan waters. However, this revolutionary stance did not last long. Because of rather than despite the presence of Communists ministers in the French government, Mitterrand was keen to demonstrate the commitment of France to NATO and, by doing so, make life for Communist ministers unbearable. The Soviet Union was enjoying a revival of unpopularity after the imposition of martial law in Poland, and the refusal of the Socialist government to condemn this in the strongest terms had proved extremely unpopular, not least among left-wing intellectuals. The Socialists were keen not to do anything that might favour the Soviets or Cubans in Central America, and in July 1984 Mitterrand received Napoleon Duarte in Paris. Even in France's Middle East policy there was a revolution, as Mitterrand brushed aside the Israeli attack on Iraq's nuclear reactor in June 1981 and paid the first state visit of a French president to Israel since its foundation in March 1982. Mitterrand did go to Moscow in June 1984, only to give a lecture on human rights in the Kremlin. And all this was against the backdrop of a wave of Americanomania in France, exemplified by college sweatshirts and fast-food chains, and by the rising popularity of Ronald Reagan, who enjoyed the confidence of 31 per cent of French people in 1982 and 40 per cent in 1984.

The shift of the Left towards the United States was paralleled by the way it learned to love the bomb. There had, in fact, long been division in the Socialist camp, those around Michel Rocard resolutely anti-nuclear, Charles Hernu and his allies in favour of the nuclear deterrent and having contacts in military circles, Chevènement and his CERES group of Jacobin patriots who were prepared to be won over. In fact the anti-nuclear door was kicked in by the Communists in the summer of 1977, when Jean Kanapa produced a report endorsing the nuclear deterrent *tous azimuts*, that is, aimed at the United States as well as the Soviet Union, as a guarantee of unqualified national sovereignty. François Mitterrand then reflected that the bomb might be France's new Maginot line, and a Socialist convention in January 1978 agreed the compromise that a government of the Left would retain the nuclear arsenal until a referendum was held on the deterrent, while campaigning for general disarmament.

Once in power the Socialists did not hesitate. Hernu was made defence minister and nothing more was said about a referendum. Mitterrand accepted the NATO plan of December 1979, agreed by Giscard, that American intermediate range nuclear weapons—Pershing IIs and Cruise missiles—would be deployed in western Europe (but not France) in order to force the Soviets to withdraw their SS20s. The benefit to France, as usual, was additional US nuclear information and technology. It also served as a lever to check any possible German drift to neutralism, and Mitterrand addressed the Bundestag on 20 January 1983 to urge acceptance of the American missiles on their soil. Whereas in Great Britain and West Germany there was powerful opposition to the arrival of these missiles, in France the anti-nuclear protest was limited and ineffective. A Comité pour le Désarmement Nucléaire en Europe (Committee for Nuclear Disarmament in Europe, or CODENE) was established in February 1982, and organized a demonstration of 30,000 Greens, PSU, and women's groups in Paris on 5 June 1982 to protest against Reagan's visit to Versailles. The main pacifist movement, the Mouvement de la Paix, however, which brought out 160,000 onto the streets of Paris on 20 June 1982 and organized a 'picnic for peace' of 300,000 on 19 June 1983, was doubly hamstrung by being an organ of the Communist party and by the Communists' acceptance of the nuclear deterrent. The fact that no American missiles were going to be located on French soil and the PCF's acceptance undermined the peace movement; but above all, most French people had swallowed the Gaullist view that their independent nuclear deterrent was a condition of French sovereignty and greatness.

The commitment of the French government and public to their deterrent was well illustrated by the *Rainbow Warrior* affair in July 1985. Atmospheric testing in the Sahara was ended in 1975 in favour of underground testing at Mururoa atoll in French Polynesia. About sixty underground tests were carried out between 1976 and 1984, and the biggest explosion, ten times that of Hiroshima, was detonated in May 1985. The South Pacific states, led by Australia and New Zealand, opposed the testing, and Greenpeace organized a demonstration in the test area. The French security services replied by sinking the Greenpeace vessel *Rainbow Warrior* in Auckland harbour on 10 July 1985, killing one of the crew. The New Zealand authorities arrested two French agents, sent them for trial, and sentenced them to ten years in prison. By contrast, the French government set up an enquiry which whitewashed the secret services and was backed up by the right-wing opposition. 'My country right or wrong' was the opinion of Giscard d'Estaing.[2] Laurent Fabius nevertheless dismissed Admiral Lacoste, the head of the security services, for failing to reveal everything to the government, and Charles Hernu resigned as defence minister in his wake. But Mitterrand flew to Mururoa in September to give his blessing to the tests, which resumed forthwith, and what was striking about the whole affair was the absence of outrage on the part of either the political class or the public.

The attitude of the French government also became clear as arms talks between Reagan and the dynamic new Soviet leader Gorbachev gathered momentum. The main concern of the French (with the British) was that their independent deterrent should not be put on the table as part of NATO's nuclear arsenal and bargained away. There was some relief when the Reykjavik summit of October 1986, agreeing to eliminate all offensive missiles in ten years, foundered on Reagan's refusal to relinquish 'Star Wars'. But Gorbachev forced the pace, proposing in the spring of 1987 that the USA and USSR first remove all intermediate weapons from Europe and then dismantle short-range nuclear weapons also. Mitterrand flew to London to confer with Mrs Thatcher, and reaffirmed France's wish that her weapons be excluded from any negotiations. The French defence minister, André Giraud, feared a 'European Munich'.[3] Gorbachev was much less popular in France than in Britain, not least because he was seen as the champion of denuclearization, and it is perhaps not surprising that, of all the Western leaders, Mitterrand was slow to condemn the military coup that temporarily overthrew Gorbachev on 19 August 1991.

Whereas de Gaulle drove American troops out of France, Mitterrand was reluctant to see the departure of American nuclear weapons from German soil, and endorsed them primarily to retain the friendship of Germany. Again, while Mitterrand followed the policy laid down by de Gaulle that France should remain outside the integrated military command of NATO, he increasingly accepted the line laid down by the United States, for the sake of the American umbrella, even when this line went contrary to France's own national interests. This was particularly clear during the Gulf war of 1991. Traditionally, France's sympathies in the Middle East were with the Arab states. Mitterrand met Colonel Gadafy on Crete in June 1985 to talk about the crisis in Chad and in April 1986 refused permission to American F111s to fly through French airspace in their way to bomb Libya. Likewise, the French enjoyed an excellent relationship with Iraq, with Mirage fighters, Exocets, surface-to-air, and anti-tank missiles following behind the nuclear plant in 1981–9. However, after Saddam Hussein invaded Kuwait in August 1990, France sent 10,000 troops to the Gulf to join 35,000 British and 540,000 Americans to enforce United Nations resolutions on Iraqi withdrawal. The French government favoured an embargo rather than war, together with an international conference on the affairs of the Middle East, including the Palestine question. When hostilities broke out in February 1991 the French defence minister, Jean-Pierre Chevènement, who had always been pro-Arab and had an old-fashioned hatred of the American 'big stick', resigned and denounced the UN Security Council as a tool of the United States, which was only interested in its oil. He was praised for his courage by the elder statesman Michel Jobert, who attacked the policy of following blindly behind the United States in the hope of a grain of greatness, falling into the trap of another Suez and destroying the trust enjoyed by France in the Arab world.

The relative independence that France had enjoyed under de Gaulle was based on an equilibrium between the superpowers. After the collapse of Soviet power, France no longer had the ability to play one superpower off against the other, and lost her room for manœuvre. When the United States insisted on action, France was bound to follow. By contrast, when France was keen to act and the United States reluctant, there was nothing France could do to prevail upon it. Two final examples may serve to illustrate this point. A clear priority for the Clinton administration was the conclusion of the Uruguay round of GATT negotiations, begun in 1986, that were designed to open the world to free trade. The French were apprehensive on two counts: the threat to the French cinema from American films, as we have seen, and the threat to French agriculture if the government were obliged to reduce subsidies to French agricultural exports. In September 1993 the Balladur government managed to cajole the Germans onto their side, then persuaded the Americans to renegotiate the deal for agriculture. The limited concessions the French were able to extract were trumpeted in France as a victory over the United States; but the farmers' unions were not convinced, and in reality the Americans had imposed the world economic order they wanted on everyone else, including the French.

At the same time, in Bosnia, where Serbs and Muslims were locked in civil war, France was involved in peace-keeping as the major provider of United Nations Protection Force. As Serb aggression against Bosnian Muslims increased, so France took a leading role in demands voiced by the international community for military action to be taken against the Serbs. The United States, on the other hand, refused to commit ground forces, and were extremely reluctant to authorize air strikes. Since Russia was not involved in the region it was of no interest to the USA, and public opinion in America would not accept American casualties in a meaningless Balkan war. Without American support, however, the involvement of NATO, in which France was also included, was impossible. Tensions between France and the United States reached one climax at the beginning of 1994, over the need to end the Serb siege of Sarajevo, and another at the end, when Serb pressure on the Muslim enclave of Bihac had to be broken. In each case, limited NATO air strikes did eventually take place against Serb positions, but France and the United Nations Protection Force virtually had to to prise action out of the Clinton administration.

Europe in the image of France

For a long time de Gaulle resisted the idea of the European Community, seeing it as a threat to French sovereignty and to French identity. But when he returned to power he accepted the Common Market that was due to come into force on 1 January 1959, seeing that France could play the role of Europe and thus acquire greater leverage on the world stage. The precondition, of course, was that France should preserve its hegemony in Europe, and that Europe should be constructed in the image of France.

In order to ensure French hegemony in Europe, de Gaulle had to lock the Federal Republic in and lock Great Britain out. The Benelux states were not in a position to offer much resistance, and de Gaulle visited Italy in June 1959 to remind the Italians that France had liberated them at Solferino exactly a century before. The key to a strong Europe was the Franco-German axis. Konrad Adenauer, the 82-year-old German chancellor, was invited to Colombey on 14 September 1958, a fortnight before the referendum on the constitution, the first of fourteen meetings until Adenauer left office in September 1963. In September 1962 de Gaulle made a triumphant tour of Germany, and declared that the union of France and Germany not only replied to the threat of the Soviet Union but established 'a bastion of power and prosperity of the same order as that constituted by the United States in the New World.'[4] The axis was further strengthened on 22 January 1963 when, a week after vetoing Britain's application to join the Six, a Franco-German treaty providing for military co-operation and cultural exchange was signed.

The Europe that emerged was that desired by France, and in particular by de Gaulle. Adenauer, divided from one half of Germany after 1961 by the Berlin Wall, looked for support against the Soviet Union and saw American protection as indispensable; de Gaulle resisted American pretensions and looked to the Soviet Union to counterbalance them. Adenauer had good relations with Great Britain; de Gaulle criticized Britain's Commonwealth and its special relationship with the United States, and looked to resurrect Napoleon's

Continental System in order to shut her out. Adenauer was very much in the mould of Monnet and Schuman, and envisaged some pooling of sovereignty in a European federation; de Gaulle, while denying that he had ever used the term *l'Europe des patries*, provoked the resignation of MRP ministers from the government in May 1962 by insisting on a *Europe des États*, and recalled France's ambassador in Brussels in June 1965 in order to oppose a scheme of majority voting in the Council of Ministers and defend the French veto. De Gaulle would never accept a Europe that called into question France as a centralized, unitary sovereign state. While he acted the bully, however, his prime minister, Georges Pompidou, presented French hegemony as quite natural. France, he told the American Club in Paris in February 1964, was 'condemned by its geography and its history to play the role of Europe'.[5]

'I do not claim that Europe should be French or speak French,' Pompidou told the Belgian newspaper *Le Soir* in May 1971. He was concerned that French would no longer be the first working language if Great Britain were allowed entry and 'then Europe would no longer be wholly European', since English was also the language of the United States.[6] Even so, it was Pompidou who lifted the French veto to British entry into Europe and bantered with the Queen at the Grand Trianon in May 1971 about how Britain must have seen the Common Market as one of those continental coalitions it had struggled against for three centuries. The referendum he had held on British entry on 23 April had received only the most lukewarm of endorsements, however, and Britain had been brought in mainly to offset the *Ostpolitik* pursued by Chancellor Willy Brandt since 1969 and realized in treaties with the Soviet Union, Poland, and the German Democratic Republic.

The closeness of the Franco-German rapport was recovered by Giscard d'Estaing, who enjoyed an excellent relationship with Chancellor Schmidt. They spoke English together, without interpreters, met informally before summits to establish a common position, often lunching at a favourite restaurant, *Au Bœuf* at Blaesheim in Alsace. Together they strengthened the institutions of the European Community, secured agreement for the European Monetary System at the Bremen summit in July 1978, introduced a six-monthly rotating chairmanship of the Council of Ministers, and introduced direct elections to the European parliament. For the European elections of June 1979 Giscard sponsored a list headed by Simone Veil, only to encounter the furious opposition of his former premier, Jacques Chirac. From his bed in the Hôpital Cochin, where he was recovering from a road accident, he unleashed a stinging attack on 'the foreign party' that was quietly scheming to dissolve the French state and identity in some supranational Europe. Giscard replied that if France wished to increase its authority it had to do so by leading the organization of Europe. Chirac was assured that his list would win 24 per cent of the vote; in the event it secured only 16 per cent. Veil's list won 27 per cent and she was duly elected president of the European parliament.

When François Mitterrand came to power in 1981 there was a loss of direction in the European Community and a loss of leadership by France. Margaret Thatcher was attacking the CAP and demanding a refund on Britain's contribution to the Community budget. Mitterrand himself floated plans for a 'social European space' at the Luxemburg summit of June 1981 that evoked only a stony silence. Some in the Socialist party demanded the isolation of the French economy and the pursuit of Keynesianism, if not socialism, in one country. Mitterrand's U-turn in March 1983 in favour of free trade and deflation was largely dictated by the realities of the international money markets and trading system, and once he had decided that France's future lay in Europe it followed that Europe should be reconstructed in France's image. Where de Gaulle had defended the centralized, unitary, sovereign state against both European federalism and French regionalism, Mitterrand was prepared to allow a more decentralized administration in France but, recognizing the limits set to French sovereignty in 1983, sought to defend that sovereignty by lodging it in a more centralized Europe, ensuring French influence in decisions made not only in economic matters but also in matters of foreign policy and defence. Roland Dumas was sent off to EEC meetings as minister for European affairs. Jacques Delors went from being French finance minister to president of the European Commission. Mitterrand himself became chairman of the Council of Ministers between January and June 1984, culminating in the Fontainebleau summit, and took the opportunity to build up support for a new, federalist Treaty of Rome.

In the event, Mitterrand's vision of a more federal Europe proved too much for many member states, and even for the Quai d'Orsay; and rather as the founding fathers of Europe settled on the Common Market after the failure of the EDC in 1954, so at Strasburg in January 1985 Jacques Delors came up with the idea of the Single European Market. Taken up in response to the growing pressure for free trade articulated by the Uruguay Round of GATT negotiations, which made it essential that Europe become a powerful, integrated economic unit, it was agreed in February 1986 that the market should come into force in 1992. In addition, the Act reactivated the idea of a single European currency, and Jacques Delors revealed his former social Catholicism rather than his socialism in his package of February 1987 which spoke of a 'social' dimension to correct the ravages of the market. The small print of the Act also stated the will of the Twelve to proceed to a European Union, including common foreign and defence policies. Nothing serious was done on this front except by the French government, which feared that West Germany was once again drifting into a neutralist position and sought to strengthen military ties with her. The revival of the Western European Union as a second pillar of NATO was proposed both to counter American withdrawal from Europe and to keep West Germany in line. A new strategy asserted that it was in the interest of France to help with the defence of Germany within Germany itself, and a Rapid Action Force (FAR) was created for use both in Germany. and in the Third World. September 1987 saw the curious coincidence of 20,000 French joining 55,000 Germans for manœuvres in Bavaria, pretending to confront a Soviet attack from Czechoslovakia, while the DDR leader Erich Honecker was warmly received by Helmut Kohl. All eyes were on Germany to see which way it would jump.

Europe in the image of Germany

François Mauriac once said that he loved Germany so much that he was delighted that there were two of them. The pulling down of the Berlin Wall in November 1989, followed by Chancellor Kohl's presentation of a ten-point plan for German unification to the Bundestag, revived many an old nightmare in France. Fears were immediately expressed of a Germany of 80 million inhabitants of massive economic power, of a Fourth German Reich hungry for *Lebensraum* that would demand a revision of its border with Poland and even covet Alsace-Lorraine, and of a Bismarckian Europe in which hegemony would be restored from France to Germany. François Mitterrand, while endorsing the German right of self-determination and declaring that there could be no opposition to a democratic and peaceful reunification of Germany, met Gorbachev in Kiev on 6 December 1989 to co-ordinate opposition to any revision of Germany's borders, and then went to the Democratic Republic to speak of the 'East German identity'. Once Chancellor Kohl, however, had done a deal with Gorbachev on the terms of German reunification, including the non-negotiability of the Oder–Neisse frontier with Poland, there was nothing that Mitterrand could do to stop it. The French would have liked to delay German reunification if they could not prevent it, but after the German elections of March 1990 had given Kohl an overwhelming mandate in both parts of Germany, it went ahead with all speed.

The response of the French government in 1990, as in 1950, was to reply to the threat of the revival of German power by seeking to lock Germany ever more tightly into a united Europe where its partners, and in particular France, could limit its freedom of action. The reconstruction of Europe, enthusiasm for which had wavered in the later 1980s, was suddenly brought back onto the agenda. For France the first priority was to deepen Europe, establishing a single currency, a European Bank to supplant the Bundesbank, which currently ran the European economy, and a common foreign and defence policy to ensure that German demographic, economic, and military resources were put at the disposal of Europe as a whole. Though there was also talk about broadening Europe, bringing in the former satellite states of Eastern Europe, this was to take second place, since their entry into the European Union would replace a French-dominated western Europe by a German-dominated *Mitteleuropa*.

The deepening of Europe was agreed under the treaty of Maastricht in December 1991 and ratified—not without controversy and conflict— by the parliaments or peoples of the Twelve. In France, as during the debates on the ratification of the EDC forty years previously, old wounds were reopened and the parties split down the middle. The Radicals, who had helped to torpedo the EDC in 1954, had learned their lesson and were now four-square behind European union, and the Communists and the National Front were solid in their opposition. On the other hand the Socialists, UDF, RPR, and the Greens were all divided, with the Jacobin Jean-Pierre Chevènement, the Vendean Philippe de Villiers, the Bonapartist Philippe Séguin, and the Franc-Comtoise Dominique Voynet respectively leading the opposition to Europe within their own parties. Fears that Europe was now dominated by Germany and that France risked losing both its autonomy and identity in a united Europe were forcefully expressed by the opponents of Maastricht. The Central Committee of the PCF denounced 'this supranational Europe dominated by Germany'.[7] Séguin, addressing the Assembly during the night of 5–6 May 1992, said:

the nation must become once again what it was: our founding principle. That implies the restoration of the State and the rehabilitation of the Republic. Nation, State and Republic, those are the means to build a Europe compatible with the idea that France has always had of itself.[8]

Supporters of the treaty argued that the French could not afford to retire to a bunker, that the French genius would continue to radiate in a united Europe, and that to bind Germany into Europe was the only way to deal with her resurgence. Mitterrand's spokeswoman, Elizabeth Guigou, pointed out that the 'Marseillaise' was sung in the Eastern bloc and the Declaration of the Rights of Man had circled the globe. Michel Rocard told supporters in Quimper that if the construction of Europe were checked, 'Germany would revert to her old historical and geographical ways. Supported by a triumphant mark, it would turn east once again, and lose interest in the future of the Continent except to impose its economic will on it.'[9] Giscard d'Estaing, asked by an old lady at the peace memorial of Caen why she should vote for Maastricht after the carnage of the Second World War, replied, 'For Franco-German reconciliation, Madame'.[10]

The Maastricht vote of 20 September 1992, the 200th anniversary of the defeat of the invading Prussians by a French revolutionary army, was a victory for Europe by only the narrowest of margins, 51 to 49 per cent. Rich, urban, and educated France voted yes; poor, rural, and uneducated France no. Centrist France voted yes, extremist France no. Alsace registered the highest proportion of yes votes, as European as it had been in 1954, Picardy the lowest. A year later a poll showed that, if the referendum were held again, the result would have gone the other way, with 56 per cent against Maastricht and only 44 in favour.[11] In the European elections of June 1994, the anti-European Mouvement des Citoyens, founded by Jean-Pierre Chevènement, secured only 3 per cent of the vote. However, the anti-European list of Philippe de Villiers, the dissident UDF deputy, won 12 per cent of the vote, reducing the score of the pro-European UDF-RPR list to 25 per cent. Dizzy with success, de Villiers finally abandoned the UDF and founded a new party, the Mouvement pour la France, which revived nineteenth-century nationalism in opposition to a federalist Europe that would serve only as a vehicle for German hegemony.

France's confusion about her European vocation was poignantly illustrated at the beginning of 1994. On the one hand she assumed the presidency of the European Union; on the other she limbered up for the French presidential elections. The first inclined her to take a positive role in building a Europe of Fifteen that now included Austria, Sweden, and Finland, and fix an agenda of 'deepening' that included a Single European Currency. The second, given growing Euroscepticism among the French electorate, encouraged politicians to take a more nationalistic view themselves. There were fears that in a Europe of up to thirty states, including most of eastern Europe, France would be marginalized on the Atlantic littoral; plans to deepen the institutions in the European 'core' were seen to be the strategy of Chancellor Kohl; either way, deepening or broadening, Germany was seen to be the victor, not France. So the rhetoric of anti-federalism and French nationalism began to take hold. Philippe de Villiers blazed a trail, abandoning the UDF and causing it to question whether it was still the vigorous pro-European party of Giscard d'Estaing. In the RPR the anti-European views of Philippe Séguin were taken up by Jacques Chirac who, campaigning for the presidency, promised a referendum on the Single European Currency. Even Édouard Balladur argued that a Europe of thirty states could not possibly be a federal Europe. François Mitterrand and stalwarts of the Socialist Party such as Michel Rocard and Elizabeth Guigou still held to the federalist vision, but without Jacques Delors as the candidate of the Left socialists were also in danger of falling prey to Euroscepticism.

From colonialism to neo-colonialism

'France has a determining role to play in preserving world peace,' wrote Michel Aurillac, Jacques Chirac's minister of co-operation, in 1987. 'She has a permanent seat on the Security Council of the United Nations, an independent nuclear deterrent and an African dimension. Were she to lose one of these pieces, she would no longer be what she is.'[12] All of these elements were, in fact, interdependent. France's colonial power ensured France a permanent seat on the Security Council in 1945, and enabled Prime Minister Bidault to have French accepted as a working language of the UN alongside English, Spanish, Russian, and Chinese. After her African colonies gained their independence in 1960 France was able to increase the size of the French-speaking bloc in the UN to twenty-two, and, adding in Morocco, Tunisia, the Lebanon, Laos, and Cambodia, managed to increase the total to a third of the UN delegates. This enabled her to match the influence exercised by Great Britain through the Commonwealth, and to sustain her position *vis-à-vis* the superpowers. Colonial possessions were also of the greatest strategic importance. The French base in Djibouti protected the oil supply routes through the Indian Ocean. Uranium from Niger went into the making of atomic bombs, which were tested first in the Sahara desert and after 1975 in French Polynesia. The possession of an independent nuclear deterrent in turn ensured the continuing occupation of a permanent seat on the Security Council.

While the preservation and promotion of France's national greatness was one enduring ambition of the French people, a second, since the Revolution, was to bring liberation and civilization to peoples less fortunate than themselves. These two ambitions were often in sharp conflict with one another, but the French managed to reconcile them in the most inspired ways. One way, as we have seen, was to impose a French view of liberation and civilization, and on French terms. When de Gaulle returned to power, the relationship of France with its colonies was revised along with the constitution. The colonial populations were given three choices in the referendum of 28 September 1958: of becoming totally assimilated with France as departments, of internal autonomy and democratic self-government within what was to be called the French Community, and of total independence, but without French aid. Only Guinea voted for immediate independence, and the French abandoned them at once, taking even their telephones. But the winds of change were blowing through the continent of Africa. The Gold Coast secured independence from Great Britain in 1957 and became Ghana. In September 1959, Mali (a federation of Senegal and Sudan) and Madagascar demanded independence within the Community. This curious status was provided for under the constitutional law of 4 June 1960, and was granted in 1960 to the thirteen African states of Cameroon, Togo, Senegal, Sudan, the Ivory Coast, Dahomey, Upper Volta, Niger, Mauretania, the Central African Republic, the French Congo, Gabon, and Chad, together with Madagascar. In 1961 it was recognized that the institutions set up to run the French Community, notably the Executive Council, Senate, and Court of Justice, no longer existed, and the Community itself evaporated.

The influence of the French in Africa did not dissolve, however; it simply took other forms. Indeed, the granting of independence, not least to Algeria in 1962, was a precondition of France developing its influence in Africa at the expense of the United States, the Soviet Union, and Great Britain. The French saw themselves as the patrons of the newly independent African states and their representative *vis-à-vis* the international community. Their votes in the United Nations were corralled to defend French interests, such as the defeat of a UN resolution to give independence to Djibouti in 1967. The Federations of West and Equatorial Africa which the Senegalese leader Senghor had defended were broken up, partly because the Ivory Coast leader, Félix Houphouët-Boigny, did not want to see his country's resources eaten up by the poor hinterland states of Chad, Sudan, Niger, and Upper Volta, and partly because the balkanization of French Africa into thirteen states with an average population of 3 million made it easier to perpetuate the influence of France. French governors left, but were often appointed the ambassadors of the new states in Paris. French civil servants stayed on to advise the new governments. Above all, co-operation agreements were signed with all the former colonies to ensure an ongoing French economic, military, and cultural presence.

None of the former colonies, except the Ivory Coast, Senegal, and Guinea, had any measure of self-sufficiency, so aid was forthcoming from France in return for privileged access to raw materials and markets. The existence of an African franc whose value was pegged to the French franc ensured that the former colonies enjoyed a hard and stable currency and that the French government had a say in their economies. Under military agreements signed with eleven of the thirteen former colonies, armies and gendarmeries about 6,000 strong in each state were built up and trained by the French. The French were allowed to keep bases at strategic points, and an intervention force, stationed in France, was held at the disposal of the new African governments. In order to perpetuate the French language and the existence of a Francophone and Francophile élite, French citizens who wished to do civilian instead of military national service were sent out, usually as teachers, while African students were encouraged to study in France. In 1970 there were 46,000 French *coopérants* in Africa, and ten years later 100,000 African students in France. In addition to all this, informal networks existed to keep the African states on a tight leash. Jacques Foccart, the Gaullist baron and the Élysée's secretary-general for African and Malagasy affairs between 1962 and 1974, acted as an informal president of Africa. He co-ordinated African affairs across the various government departments, addressed African leaders by the familiar 'tu', made or broke their careers, and had at his disposal a network of secret agents the existence of which was always firmly denied.

Giscard d'Estaing dismissed Foccart, only to give the job to his assistant, René Journiac. His aim was to ensure that the French grip on Africa was broad and strong enough to resist the influence both of the Anglo-Saxons and of the Soviets, who were supplying Colonel Gadafy with arms and had sponsored the Angolan revolution of 1975. He regularized the Franco-African summits begun by Pompidou in 1973, and included in them the former Belgian colonies of Zaïre, Rwanda, and Burundi, and the former Portuguese colonies of Guinea-Bissau, the Cape Verde Islands, and São Tomé. At the Kigali (Rwanda) summit of May 1979, he even floated the idea of bringing Arab countries into the fold. France already had the habit of putting its military weight behind sympathetic African leaders, for example supporting President M'Ba of Gabon against a military coup in 1964. Under Giscard this support for dictators became explicit. In 1977 he supported President Mobutu of Zaïre against Angola-backed Congolese rebels, and the following year supported Hissan Habré of Chad against rebels supported by Libya. Most extraordinary was his patronage of Jean-Bedel Bokassa, providing 4 million francs for his Napoleonic coronation as emperor of the Central African Republic on 4 December 1977, receiving lavish gifts of diamonds in exchange, only to dump him in 1979 when it became clear that he was receiving support from Libya.

After the Socialists won power in 1981, a full-frontal attack on this neo-colonialism was led by the young minister of co-operation, Jean-Pierre Cot. The son of Pierre Cot, who had been Radical air minister in the Popular Front government but moved closer to the PCF after the war as a result of his involvement in the peace movement, he was determined to construct a more egalitarian and more principled relationship with the Third World. Aid to developing countries, which had stood at 1.4 per cent of GNP in 1960, had fallen to 0.67 per cent in 1968 and 0.34 per cent in 1979. Mitterrand had promised to raise it to 0.7 per cent, and by 1983 it had returned to 0.5 per cent. Cot sought to spread the aid more fairly, reducing the quasi-monopoly of black Africa and sending more, for example, to Latin America. He wanted to ensure that Africa developed balanced economies, not just economies that suited the economic needs of France, and that the massive problem of Third World debt was tackled. He channelled more aid to non-governmental projects and less to governments that simply used it for prestige projects and to build up their bureaucracies. He asserted the primacy of the rights of man, criticized France's strategy of supporting dictators, and received opposition leaders in Paris. A scandal broke out when he refused to meet Sekou Touré of Guinea, who was alleged to have killed many of his opponents, when he was invited by Mitterrand to Paris in September 1982. Cot was obliged to attend the ceremony nevertheless, and resigned on 8 December.

Cot's position had in any case become untenable. The old networks persisted, and heads of state communicated behind his back with Guy Penne, Mitterrand's adviser on African affairs at the Élysée. Soon things were back to their old corrupt and repressive ways. The new Socialist minister of co-operation, Christian Nucci, and the head of his private office set up an association called Carrefour du Développement to alert public opinion to Third World problems and to finance the next Franco-African summit. After the Socialists left power in 1986, it became clear that money had been siphoned off to build up Nucci's power base at Beaurepaire in the Isère and to buy a château. Nucci was sent to the Haute Cour de Justice but granted amnesty under a law of December 1989; the head of his private office, Yves Chalier, who fled to Rio, was sent for trial in 1992. Meanwhile, in 1983, Mitterrand became involved in sending French troops to Chad, to defend French interests there against Libyan-backed forces. The problem of coping with such endemic instability led to the formation by Charles Hernu of a Force d'Action Rapide (Rapid Action Force, or FAR), 47,000 strong, on full alert in France to intervene either in Africa or in Germany. The number of African officers trained in France shot up during the 1980s, although arms supplies were diverted to the Maghreb and the Middle East, because they were regarded as less volatile.

While relations with some dictators, like Mobutu of Zaïre, grew cooler after 1990, those with other dictators, such as Habyarimana of Rwanda and his Hutu regime, grew warmer. After an invasion by the Uganda-based Rwanda Patriotic Front (RPF), composed of the formerly dominant Tutsi tribe, in October 1990, the French stepped up military aid to Habyarimana and helped to train his forces. Habyarimana was assassi-nated in April 1994 and the Hutus began to massacre their Tutsi opponents. The French, desperate to act independently and to send in more troops, claimed that their aim was humanitarian, to prevent geno-cide. In fact, they were keen to prevent the rebel RPF from seizing Kigali and destroying French influence in the region. Kigali fell to the rebels in July 1994 and the French, isolated in the UN, received a green light only to set up a 'secure humanitarian zone' on the Zaïrean border for Hutus fleeing the new regime. French troops withdrew in August 1994, having failed both to convince the international community that their mission was humanitarian rather than military and to protect the interests and ascend-ancy of France in central Africa. They paid the price for their support of reactionary and repressive regimes, and yet their pretensions in Africa were in no sense deflated. Addressing his last African summit in Biarritz in November 1994, President Mitterrand, flanked by the presidents of Gabon, Togo, and Zaïre, announced that 'France must hold to her road and refuse to reduce her African ambition . . . France would no longer be quite the same in the world if she abandoned her presence in Africa.'[13]

Defending the last colonies: New Caledonia

There were some parts of the French Union—the Overseas Depart-ments and Territories or DOM-TOMs—to which the choice of inde-

pendence offered in 1958 did not apply. In the first place, there were
overseas departments of France that were governed directly from Paris:
Algeria, of course, but also Martinique and Guadeloupe in the French
West Indies, Guiana in South America, and Réunion in the Indian
Ocean. Secondly, there were overseas territories, at strategic points
around the globe, which were granted full or partial independence only
with the greatest of reluctance. These included the islands of Saint-
Pierre and Miquelon off Newfoundland, French Somaliland or Djibouti,
which was most unruly and was granted independence after a referendum in June 1977, and the Comores islands in the Indian Ocean, which
became independent after a referendum in December 1974. In the
South Pacific there were the islands of Wallis and Futuna, which gave
no trouble, the New Hebrides, ruled jointly with Great Britain and
called Vanuatu after independence in 1980, French Polynesia (including
Tahiti), which was granted internal autonomy in 1984 but not more,
because of French nuclear testing, and New Caledonia.

New Caledonia, French since 1853, became a boom colony in the
1970s when the nickel mined there came into great demand for steel and
armaments. There was massive immigration from other Pacific islands
and of *pieds-noirs*, so that the European or Caldoche population
outstripped the indigenous Melanesians, reduced to 42 per cent of the
population in 1976. In response to the arrival of a Socialist government
in France and moves towards independence in other parts of the
Pacific, a Front de Libération Nationale Kanake et Socialiste (Kanak
National and Socialist Liberation Front, or FLNKS) was formed in
1984. This provoked the formation by the Caldoches of a Rassemblement pour la Calédonie dans la République (Union for Caledonia in
the Republic, or RPCR), which won the elections to the territorial
assembly in November 1984. The FLNKS, which had boycotted the
elections, established a provisional government of an independent
Kanaky under the presidency of Jean-Marie Tjibaou. The new high
commissioner, Edgar Pisani, proposed 'independence in association
with France', and Prime Minister Fabius promised a referendum on
independence and set up four regional councils, control of three of
which was won by the independence movement in September 1985.

The RPR manifesto for the 1986 elections was full of rhetoric about
France's liberating and civilizing mission. In office, however, the Chirac
government responded to the RPCR and was determined to retain full
control of the South Pacific. Power was removed from the regional
councils and concentrated in the high commissioner and RPCR-
controlled territorial council, while the French military presence was

increased to 6,000. The referendum on independence, held in September 1987, was boycotted by the FLNKS, and 57 per cent of the vote endorsed a continuing relationship with France. Bernard Pons, the Minister for DOM-TOMs, introduced a new statute to reduce FLNKS control to one out of three regions and clamped down on the independence movement. In desperation, the Kanaks kidnapped 27 gendarmes, four of whom were killed, two days before the first round of the presidential elections in April 1988, and imprisoned them in a cave on Ouvéa island.

At moments like this, the demands of the French liberating and civilizing mission came into direct conflict with the demands of French greatness and honour. Jacques Chirac, premier and candidate in the final round of the presidential elections, did not hesitate: 300 crack troops were sent in to liberate the hostages on 5 May, killing 21 Kanak freedom fighters and losing two men in the process. 'It was a matter in which the honour of France was at stake,' asserted Bernard Pons, 'a matter that concerned the honour of the French army and the honour of the gendarmerie.'[14] Not all French people, however, felt the same way. Demonstrators marched to the Hôtel de Ville in Paris on 5 May shouting 'Independence for Kanaky' and 'Pons, Pasqua, assassins'. In spite of his strong-arm tactics, Chirac was soundly beaten in the presidential election on 8 May.

Michel Rocard, the new prime minister, brought Tjibao and the RPCR leader Jacques Lafleur to Paris for talks. The Matignon agreement of June 1988 provided for direct rule by France for a year, a reform that would give the independence movement control of two out of four regional councils, the economic and educational development of Melanesian regions, and a new referendum on independence in 1998. This was approved by the National Assembly, then put to the French people in a referendum in November 1988. It was approved by 57 per cent to 43 per cent, but in fact a record 63 per cent of the electorate abstained. Clearly, what happened to small French possessions in the South Pacific was only of limited interest to French citizens.

Last incarnation of the civilizing mission: Francophonie

In one sphere the French did try genuinely to develop their liberating and civilizing mission unencumbered by economic and military considerations, if not totally free of the *arrière-pensée* of French greatness.

That was in the gathering of those French nations who spoke French primarily or secondarily, to ensure the radiation of French language and culture and to draw together a world-wide community whose ties were linguistic and cultural rather than based on history, economics, or power. Of course, the enterprise was not simply a naïve exercise in cultural contacts. Cultural contacts were inevitably a channel for aid and trade, and aid and trade a means of broadening French influence in the world beyond the historic Empire or the reach of military force. The defence of the French language was an attempt to meet the challenge of other languages, notably English, Arabic, and Creole, and to defend the political influence that undoubtedly went with linguistic supremacy. The key to Francophonie was the axis between France and Canada, which had a large French-speaking population, and was thus a response both to American hegemony and to the British Commonwealth.

The initiative for Francophonie came initially not from the French but from the leaders of newly independent Francophone Africa, Léopold Sédar Senghor of Senegal, who was a distinguished poet as well as a politician, Habib Bourguiba of Tunisia, and Hamani Diori of Niger. They argued that the use of the French language produced a spiritual and intellectual community which needed to be structured and organized for the mutual benefit of all its members. Having themselves long been deputies in the French National Assembly, they set up an Association Internationale des Parlementaires de Langue Française in Luxemburg in 1967, with delegates from twenty-two countries, which agreed to defend French language, culture, and civilization in countries which were wholly or partly French-speaking. They were instrumental in creating the Agence de Coopération Culturelle et Technique at Niamey (Niger) in 1969. The meeting was attended by the French minister of culture, André Malraux, and the Agency, funded mostly by France, Belgium, and Canada, facilitated cultural and technical contacts between states, assisting their economic and educational development.

The Canadian axis was fundamental for the emergence of Francophonie as a serious force in the world. When de Gaulle declared 'Long live free Quebec!' in July 1967, Montreal was the largest French-speaking city in the world outside Paris, and Quebec, three times the size of France, was five-sixths French-speaking. The advent of the pro-independence Parti Québecois to power in November 1975 was a mixed blessing for Francophonie, however, as it was regarded as a threat by the federal government in Ottawa. The solution was to bring Canada

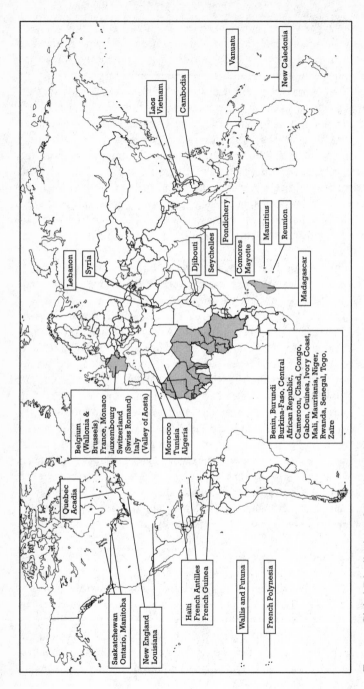

FIG. 10. Global Distribution of French-Speaking Communities.
Source: William W. Bostock, *Francophonie* (Melbourne, River Seine Publications, 1986).

as a whole, rather than just Quebec, into the fold of Francophonie, although the dominant position of France within the movement was henceforth at risk.

Once they had taken an interest in the concept, the French were quick to coordinate and develop it. A Comité de la Francophonie, under the French prime minister, was set up in November 1973 by President Pompidou, who also launched the Franco-African summits, discussed above. In March 1984 Mitterrand set up the Haut Conseil de la Francophonie, under his chairmanship and with Senghor his deputy, bringing academics, journalists, media men, film-makers, and UNESCO chiefs to reflect on the nature and ideals of Francophonie and to suggest new developments. The most important was the summit of heads of state and government of countries using French, which met in Paris and Versailles in 1986, Quebec in 1987, Dakar in 1989, Paris again in 1991, and Mauritius in 1993. The membership of these summits extended far beyond the range of the former French empire. At the first Paris summit there were forty-two states, including France, Belgium, Switzerland, Luxemburg, and Monaco from Europe; Canada, Quebec, New Brunswick from North America, with a special guest from Louisiana; the Dominican Republic, Haiti, and Saint-Lucia from the Caribbean; Tunisia, Morocco, and Mauretania (but not Algeria) from the Maghreb; nineteen Black African states, including Madagascar; Egypt and Lebanon from the Middle East; observers from Vietnam and Laos; Mauritius, the Comores, and Seychelles from the Indian Ocean, and Vanuatu from the Pacific. At the second Paris summit, fifty members were announced, including Cambodia, Romania, Bulgaria and the tiny communities of the Val d'Aoste and the 'Francos' of New England, which brought Francophonie up to the same number of participants as the Commonwealth, though the world had 370 million English speakers as against 130 million speakers of French. New technology, meanwhile, including satellite and cable television, which made possible an audiovisual university, helped to give substance to this international community.

Francophonie was not, however, a community without problems. There was first the legacy of French colonial rule, so that Vanuatu, which attended the first summit in 1986, refused to attend the second in 1987, because of French treatment of New Caledonia. Second, the commitment of Francophonie to human rights and democratic values made the participation of dictators like Marshal Mobutu of Zaïre somewhat embarrassing. Plans to hold the fourth summit in Zaïre were abandoned, Mobutu did not attend the summit when it was eventually held in Paris, and Mitterrand only reluctantly agreed to meet him at the end of the Mauritius summit.

Third, France found itself sharply challenged for the leadership of Francophonie by Canada. The Canadian premier, Brian Mulroney, brought off a coup at the Quebec summit by announcing the cancellation of the debts of seven African countries and a million dollars of aid to Lebanon. The French feared Canadian penetration of African markets under the cover of Francophonie, and Mitterrand responded at the Dakar summit by announcing the cancellation of nearly half the debt of thirty-five African states. Fourth, though Francophonie included the Arab states of Africa, the challenge of the Arab language, the rise of Islamic fundamentalism, and the existence of another organizing centre in the Arab League together weakened whatever grip Francophonie had. Algeria, pursuing a policy of Arabization and under pressure from the Islamic Salvation Front, remained outside Francophonie. The Dakar summit of May 1989 was undermined by an Arab League summit held at Casablanca at the same time. Hosted by King Hassan of Morocco, it was concerned to bring Egypt, suspended from the League in 1982 for recognizing Israel's right to exist as a state, back into the fold. The Dakar summit was also boycotted by Mauretania, its Islamic regime attacked by Senegal for persecuting the minority black population there.

Fifth and last, though the Mauritius summit of 1993 declared its renewed resolve to defend the French language worldwide, France itself was reduced to using legal means to protect French within its own borders. A law sponsored in the wake of the Mauritius summit by the minister of culture and Francophonie, Jacques Toubon, required the use of French in scientific publications and conferences, and imposed fines of up to 10,000 francs for the use of foreign terms in advertising where French equivalents existed. Not only was this law of June 1994 laughed out of court by the French press and media, which had evolved their own *franglais* over the decades, but the articles of the law banning the use of foreign terms were overruled by the Conseil Constitutionnel on 29 July 1994, on the grounds that they violated the right of French people to communicate their thoughts and opinions freely under article 11 of the Declaration of the Rights of Man and of the Citizen. The civilizing mission and liberating mission of France here clashed head on, and on this occasion the former had to give way.

Conclusion
The Challenges Facing France

The great achievement of France in the last fifty years has been to rise to the international challenges of the late twentieth century while preserving a very specific French identity. She has adapted to the modern world while remaining faithful to her past. The passionate commitment to certain principles and certain ideas, however, while contributing to the resilience of the French, also involves the risk of a certain inflexibility, even fragility, that may reduce her ability to adapt to challenges ahead. It may be that greater flexibility and imagination will be necessary to defend that identity in the long run.

France has struggled to protect her competiveness in the cut-throat markets of the world. She has developed a modern, high-tech sector which is capable of penetrating and holding its own in world markets. On the other hand, her most successful export is armaments, and given that much of the traditional industrial base has been shut down, France is reliant on tourism and agricultural exports to bridge the balance of payments deficit. France's Great Leap Forward was an achievement of the planning of the post-war era; measures will have to be taken to prevent her sliding again into an economic backwater.

The tension between modernization and traditionalism is nowhere sharper than in agriculture. France is driven by international competition to privilege the largest, most modern, and capital-intensive farms and to drive smaller, less efficient farmers off the land. Yet as, of all people, a Communist deputy from the declining industrial department of the Pas-de-Calais put it in 1993, 'Rural society bears the great values of solidarity and humanity which constitute the coherence and identity of France.'[1] One of the great issues of the end of the century is whether the countryside should be turned into a rural factory, managed by at most a quarter of a million farmers, mostly living in towns, while the countryside fills up with commuters, retired people, and foreigners, or whether the vestiges of a peasantry should be subsidized, not for its economic value but in order to perpetuate the myth of rural France and the values associated with it. The political influence exercised by the peasantry, out of all proportion to their numbers, is in large part explained by the fact that they have not only been tillers of the soil but bearers of that myth. It may be that the final liquidation of the peasantry has to be accepted, and other ways sought to sustain the myth and values of rural France.

Another challenge, common to all Western societies, is the extent to which economic indicators such as the balance of payments, the rate of inflation, and the level of public borrowing take precedence over social issues such as unemployment and poverty. After a brief flirtation with Keynesian reflation, French governments accepted the discipline of the international capitalist economy and stuck to deflation. But while this has brought inflation, government spending, and the trade deficit under control, it established a seemingly immovable level of unemployment at 10 or 12 per cent of the working population, and created a class of 'new poor' described by social theorists as 'marginal' to society. The French do not have the luxury of arguing that a change of government would solve the problem, for conservative and socialist governments have pursued essentially the same policies and bear equal responsibility. The political class and voting public as a whole may decide to accept high levels of unemployment and poverty as necessary evils of late capitalist society, but if they do, the revolutionary values of Liberty, Equality, and Fraternity may be irredeemably tarnished.

This leads to the wider issue of politics and politicians. The French seem to have surmounted the perennial crises of state that dogged their history as they lurched from chronic political instability to revolution and *coup d'état*. The 'French exceptionalism' has ended, as France has now developed a stable constitution, pluralist democracy, national consensus, and the convergence of the major parties on the centre ground of politics. And yet the French public has rarely been more disillusioned with its politicians and political parties. Under the Fourth Republic, hostility to politicians and political instability went hand in hand; now there is political stability, but still hostility to politicians. The political class is generally seen to be both incompetent and corrupt, unable to tackle pressing economic and social problems, while trading principle for opportunity in furthering their own careers and abusing relationships with the business world to feather their own nests.

The issue of corruption, on a scale greater than that of the Panama scandal of the 1890s or the Stavisky scandal of the 1930s, has done more than anything to discredit politicians and increase political apathy and disillusionment. At least it seems that, in France, guilty politicians resign and some are even brought to justice. Legislation has also been brought forward to end the financing of political parties by business, outlaw favouritism in the granting of public contacts, and oblige politicians to declare their interests. On the other hand, politicians have been markedly reluctant to reduce pluralism in office-holding, which is largely responsible for establishing them as a caste. In addition, it is obvious that most of the old political parties are in urgent need of renewal, in terms of both ideology and personnel. The old faces have been there too long, the party bureaucracies are too rigid, and ideology has withered away. The politicians themselves accept this: at the end of 1994 Michel Rocard denounced his own party as 'a heap of ruins.'[2]

It is not that there is a dictator waiting in the wings with a broom to sweep aside corrupt and clean politicians alike. But without reform and renewal three possibilities emerge. The first is that votes may shift to the National Front, which has been campaining hard against corruption, helping the party over the threshold of 15 per cent that it has found it impossible to cross. A second is that influence will shift to political mavericks such as Philippe de Villiers and Bernard Tapie, who are seen to be charismatic individuals independent of the party machines, and having a direct rapport with the electorate. A third is that the electorate, especially younger voters, will abandon traditional catch-all political parties for single-issue movements, and evolve a new form of politics related to their own subcultures. The political class will have no purchase on them, and will be seen to be increasingly redundant.

A further challenge for politicians and, indeed, for the French in general is to take a hard look at the founding myths of the One and Invisible Republic. While contributing to the development of a strong state and united nation, they encourage uniformity, conformity, and intolerance that may in the end fragment French state and society. The governing class, defended as a meritocracy, resembles nothing more than the Chinese mandarinate. Recruitment by the royal road of the *grandes écoles* requires long training and formal education that are accessible only to a certain bourgeoisie, which thus perpetuates itself in the public and private bureaucracies. This élitist system will have to be opened up to new blood and new ideas if it is going to meet the challenges of the next century.

Feminism, as we have seen, has made only limited progress in France, and it may now be that the time for reform has passed. The marginalization of feminism in the dominant discourse, however, is echoed in a wider tendency to refuse self-expression and group identity to gays and other sexual minorities. Political correctness is not much appreciated by the French, there is a constant fear of appearing ridiculous, and meanwhile the ravages of AIDS are proportionately far higher than in Great Britain and the United States. The paradox between French individuality and French intolerance needs to be explored, and the boundaries between what is socially acceptable and what is not need to be relaxed.

This intolerance is evident, more importantly, in issues of religion and ethnicity. The French state has proclaimed itself secular and has required religious groups to confine their religious practice to the private domain. Ethnic minorities exist in France, but they have no rights or representation as communities *vis-à-vis* the state. In order to acquire French nationality, individuals have to meet criteria of parentage and residence and now submit a request to join the national community, which presumably allows the authorities to reject the request. Assimilation of the French language and civilization is a condition of exercising political rights as a French citizen.

This strategy has had the remarkable effect of forging a French nation from many diverse peoples. But it has been based on the premiss that unity requires uniformity, and has resolutely refused to tolerate difference. For two centuries this may have been a strength: the threat to the state from Roman Catholicism has been dealt with; Bretons, Alsatians, Corsicans have been forged into one nation; and ethnic minorities from southern and eastern Europe assimilated. What has proved more difficult to deal with is first, the Islamic religion, and second, the Arabs from North Africa. The French have stuck to their guns, banning the Muslim veil from state schools, tightening up their nationality laws, expelling illegal immigrants, insisting on the assimilation of ethnic minorities, refusing any drift towards a multiethnic or multicultural society on what they call the Anglo-Saxon model. And yet France is clearly, in fact, a multiethnic society. The great challenge in this respect, at the end of the twentieth century, is whether the French will accept any relaxion of the assimilationist model, recognize the right of other communities to retain their differences, and accord them rights as such, starting with the right of non-European residents to vote in local elections. For French intolerance has two faces: one is the National Front, the other the Jacobin insistence on a homogeneous nation-state. The former holds that ethnic minorities cannot be assimilated and must be expelled; the latter that they need not be expelled but must be integrated. The danger is that minorities, insisting on their differences, and feeling themselves persecuted, will reject the French nation, develop their separate identity through Islamic fundamentalism, and feel greater loyalty to Algeria or Iran than to France.

The position of France herself in the world must also be reviewed. 'The role of France is to retain its rank,' said François Mitterrand in

1989. 'It is by . . . the will to retain its rank', echoed Foreign Minister Alain Juppé in 1994, 'that France can affirm itself as she wants to be: a great power.'[3] This rhetoric about national greatness and world-power status has long outlived its useful life. It compensates for the absence of greatness and declining international status, and can only create false expectations and eventual disappointment. It creates an unnecessary obligation on France to cut a magnificent figure on the world stage which her resources are in no way large enough to sustain. The first challenge in this respect is, therefore, for France to accept the fact that she is a medium-sized power, and to tailor her ambitions and pretensions in the light of that.

In one sphere, the French have become aware of their relative smallness, and that is in Europe. It is now clear that, though the French entered into the building of Europe to contain Germany, Germany and not France is now the dominant power in Europe. The reaction of enlightened French statesmen has been to secure the Franco-German axis as the basis of the European Union, and to take a positive line in deepening the Union in order to increase suprantional controls over Germany. The referendum on the Maastricht treaty in 1992 endorsed this strategy by a whisker. Since then, however, a feeling of powerless-ness has grown in France, and with it the sense that France should have no more truck with federalism and fall back on a sound, old-fashioned, anti-German nationalism. The view that federalism can secure French national interests is now giving way to the view that federalism is a German idea and a threat to the French nation-state. The challenge in this respect is whether France can shake off this 'little French' nation-alism, and recover its confidence in the European order. If it is fearful of Germany there is ample scope to develop ties with the United Kingdom, for example in matters of military co-operation. But, in the end, to abandon the Franco-German axis will harm only itself.

If France needs to be more confident in her strength in Europe, as a former colonial power she needs to become more reticent. Her repeated military interventions in Africa have looked like antiquated colonial wars. Support for African dictators and a refusal to let go of her vestigial colonies sits ill with the long-proclaimed liberating mission of the French. The challenge here is to accept that the way forward lies through the civilizing mission, and that Francophonie is beautifully adapted for this purpose. Aid and trade will follow the French language and civilization far more consistently than they will follow the flag.

While the French government and intellectuals constantly worry about the threat to the French language and culture, it would seem that in this respect their fears are not wholly substantiated. Intellectuals may be in crisis, but this reflects only the multiplication and diversification of expertises. French culture is under threat from American culture, but the French novel, French poetry, the French *chanson*, the French *bande dessinée*, and above all the French cinema all flourish. EuroDisney may be a scar on the French landscape, but the idea has been successfully converted to support French popular culture in the Parc Astérix, near Roissy airport, Schtroumpfland, in Lorraine, and the Futuroscope of Poitiers. No doubt the French-speaking world has to be cultivated and sustained by the institutions on Francophonie, but 130 million speakers of French world-wide is no small public. Above all, though there is a constant debate about what French cultural policy should be, at least there are cultural policies, at least there is public patronage for the arts, both national and local, at least the French remain self-conscious about their creative genius.

One final challenge: the French must take responsibility for their own history. In the first place, this means collectively accepting responsibility of how the French behaved under the Vichy regime and German Occupation. To blame everything on a few war criminals is to acquit everyone else; to collapse the experience of Vichy into that of the Holocaust is to conceal its other crimes; to question the integrity of the Resistance is to promote the cause of Pétainism. But coming to terms with their own history may be understood in another sense. *Grandeur*, Jacobin centralism, Gaullism, anticlericalism, the civilizing mission are all elements that have defined French history and French identity. Their remain part of that history and that identity, but they must be reconsidered critically and with discrimination if they are not to become burdens that inhibit the French from adapting imaginatively to the challenges of the next century.

Notes

Chapter 1 (pages 5–29)

1. Quoted in Klaus Manfrass and Jean-Pierre Rioux, *France-Allemagne 1944–1947. Akten des Deutsch-Französischen Historikerkolloquiums*, Baden-Baden, 2–5 Dec. 1986, Cahiers de l'Institut d'Histoire du Temps Présent, 13–14 (Dec. 1989–Jan. 1990), 15, 38.
2. Irvin M. Wall, *The United States and the Making of Postwar France, 1945–1954* (Cambridge, CUP, 1991), 76.
3. Georges Soria, *La France deviendra-t-elle une colonie américaine?* (Paris, Éditions du Pavillon, 1948).
4. Quoted in Laurent Greilsamer, *Hubert Beuve-Méry, 1902–1989* (Paris, Fayard, 1990), 335, and Richard Kuisel, *Seducing the French. The Dilemma of Americanization* (Berkeley, U. of California Press, 1993), 65.
5. René Girault and Robert Frank (eds.), *La Puissance française en question, 1945–1949* (Paris, Publications de la Sorbonne, 1988), 24, 107.
6. *Journal officiel de la République française. Débats parlementaires. Assemblée Nationale*, 11 June 1948, 3461–2, 3485, 3487–8.
7. Ibid. 29 Aug. 1954, 4468.
8. Charles-Robert Ageron, 'La Survivance d'un mythe. La Puissance par l'empire coloniale (1944–1947)', in Girault and Frank, *La Puissance française*, 32.
9. Paul Isouart, 'Les Aspects politiques, constitutionnels et administratifs des recommandations', in Institut Charles de Gaulle/Institut d'Histoire du Temps Présent, *Brazzaville, janvier–février 1944. Aux sources de la décolonisation* (Paris, Plon, 1988), 81.
10. John W. Young, *The Cold War and the Western Alliance, 1944–1949* (Leicester, Leicester UP, 1990), 56.
11. Alexander Werth, *The Strange Story of Pierre Mendès France and the Conflict over French North Africa* (London, Barrie, 1957), 148.
12. Pierre Mendès France, *Dire la vérité. Causeries du Samedi, juin 1954–février 1955* (Paris, Julliard, 1955), 60.
13. Jacques Soustelle, *Le Drame algérien et la décadence française* (Paris, Plon, 1957), 68.
14. David L. Schalk, *War and the Ivory Tower* (New York, OUP, 1991), 65.
15. Jean-Pierre Rioux, *La Guerre d'Algérie et les français* (Paris, Fayard, 1990), 228.
16. *Le Monde*, 21 Apr. 1956, cited by Jean-François Sirinelli, 'Guerre d'Algérie, guerre des pétitions?', in J.-P. Rioux and J.-F. Sirinelli (eds), *La Guerre d'Algérie et les intellectuels français* (Cahiers de l'Institut d'Histoire du Temps Présent, 10, (Nov. 1988), 189–90.

17. Pierre Vidal-Naquet, *La Torture dans la République* (Paris, La Découverte/Maspéro, 1983), 67.

18. Jean-Pierre Vittori, *Nous, les appelés d'Algérie* (Paris, Stock, 1977), 121, 286–8.

19. See below, p. 75.

Chapter 2 (*pages 30–55*)

1. Quoted in Grégoire Madjarian, *Conflits, pouvoirs et société à la libération* (Paris, Union Générale des Éditions, 1980), 195.

2. Jacques Duclos, *Batailles pour la République* (Paris, Éditions Sociales, 1947), 23.

3. Jacquier-Bruère [pseud. of Michel Debré], *Refaire la France* (Paris, Plon, 1945), 122.

4. Vincent Auriol, *Journal du Septennat* vi (Paris, Armand Colin, 1978), 15.

5. Charles de Gaulle, *Discours et messages* (Paris, Plon, 1970), ii. 582.

6. Pierre Mendès France, *Œuvres complètes*, ii: *Une politique de l'économie, 1943–1954* (Paris, Gallimard, 1985), 479.

7. Pierre Poujade, *J'ai choisi le combat* (Saint-Céré, Société Générale d'Éditions et des Publications, 1955), 29.

8. De Gaulle, *Discours et messages*, iii. 10.

9. *L'Humanité*, 28, 30, 31 May 1958.

10. De Gaulle, *Discours et messages*, iii. 44, 64, 166; iv. 167–8.

11. Quoted by Jacques Capdevieille and René Mouriaux, *L'Entre-deux de la modernité. Histoire de trente ans* (Paris, FNSP, 1988), 57.

12. Gaston Monnerville, *Vingt-deux ans de présidence* (Paris, Plon, 1980), 215.

13. Michel Debré, *Trois Républiques pour la France. Mémoires*, ii: *1946–1958* (Paris, Albin Michel, 1988), 411.

14. Alain de Boissieu, *Pour servir le Général, 1946–1970* (Paris, Plon, 1982), 188.

Chapter 3 (*pages 56–78*)

1. Pétain, broadcast of 17 June 1941 in *Actes et écrits* (Paris, Flammarion, 1974), 551–2.

2. Charles de Gaulle, broadcast of 14 Oct. 1944 in *Discours et messages*, i. 455.

3. Herbert Lottman, *The People's Anger* (London, Hutchinson, 1986), 278–9.

4. *Journal officiel. Assemblée Nationale. Débats parlementaires*, 21 Oct. 1952, 4248.

5. Henry Rousso, 'Vichy, le grand fossé', *Vingtième siècle*, 5 (Jan.–Mar. 1985), 76.

6. Henry Rousso, *Le Syndrome de Vichy, 1944–198 . . .* (Paris, Seuil, 1987), 129–30.

7. Press conference of 21 Sept. 1972, cited in René Rémond, *Paul Touvier et l'Église* (Paris, Fayard, 1992), 381.

8. De Gaulle, *Discours et messages*, v. 232.

9. Serge Klarsfeld, *Le Mémorial de la déportation des juifs de France* (Paris, the author, 1978).

10. *Le Monde*, 28 Dec. 1978.

11. Ibid. 18 July 1992.

12. Pascal Perrineau, 'Le Front National, 1972–92', in Michel Winock (ed.), *Histoire de l'Extrême Droite en France* (Paris, Seuil, 1993), 276.

13. Georges Kantin and Gilles Manceron (eds.), *Les Échos de la mémoire. Tabous et enseignement de la seconde guerre mondiale* (Paris, Le Monde-Éditions, 1991), 24.

14. *Le Monde*, 23 Mar. 1992.

15. *Esprit*, 198 (Jan. 1994).

16. Alain Finkielkraut, *La Mémoire vaine du crime contre l'humanité* (Paris, Gallimard, 1989), 78.

17. *National-Hebdo*, 16 Apr. 1992.

18. *Libération*, 15 Apr. 1992.

19. *Le Monde*, 20 Apr. 1994.

20. *Libération*, 14 June 1994.

21. *Le Monde*, 19, 23 April 1994.

22. Ibid. 14 Sept. 1994.

Chapter 4 (*pages 79–109*)

1. Jean Fourastié, *Les Trente glorieuses ou la Révolution invisible* (Paris, Fayard, 1979).

2. Dominique Merllié and Jean Prévot, *La Mobilité sociale* (Paris, La Découverte, 1991), 53.

3. Jean-Paul Flamand, *Loger le peuple. Essai sur l'histoire du logement social* (Paris, La Découverte, 1989), 295–6.

4. Jean Labbens, *Le Quart-Monde* (Pierrelaye, Éditions Science et Service, 1969).

5. André Gueslin, *Nouvelle histoire économique de la France contemporaine*, iv: *L'Économie ouverte, 1948–1990* (Paris, La Découverte, 1989), 7, 24.

6. Christian Baudelot, Roger Establet, and Jacques Malemort, *La Petite Bourgeoisie en France* (Paris, Maspéro, 1974), 93–4.

7. Gérard Noiriel, *Workers in French Society in the 19th and 20th Centuries* (New York, Berg, 1990), 181–95.

8. Jacques Frémontier, *La Forteresse ouvrière: Renault* (Paris, Fayard, 1971), 109; Pierre Naville, J.-P. Bardou, P. Brachet, and C. Lévy, *L'État entrepreneur. Le Cas de la régie Renault* (Paris, Anthropos, 1971), 183.

9. Henri Mendras and Alistair Cole, *Social Change in Modern France* (Cambridge/Paris, CUP/Maison des Sciences de l'Homme, 1991), 32.

10. John S. Ambler, *The French Welfare State. Surviving Social and Ideological Change* (New York: New York University Press, 1991), 61.

11. Christian de Montlibert, *Crise économique et conflits sociaux dans la Lorraine sidérurgique* (Paris, L'Harmattan, 1989), 169.

12. J.-P. Durand and F.-X. Merrien, *Sortie de siècle. La France en mutation* (Paris, Vigot, 1991), 111–13.
13. Jean-Pierre Terrail, *Destins ouvriers. La Fin d'une classe?* (Paris, PUF, 1990), 168–9, 197–9.
14. Patrick Champagne, 'La Manifestation. Production de l'évènement politique', *Actes de la recherche en sciences sociales*, 52–3 (June 1984), 19–41.
15. *Le Monde des débats*, Nov. 1992.
16. *Le Monde*, 25 Feb. 1994.
17. 'Le Revenu minimum d'insertion', *Économie et statistique*, 252 (Mar. 1992), 16.
18. Serge Paugam, *La Disqualification sociale. Essai sur la nouvelle pauvreté* (Paris, PUF, 1991).

Chapter 5 (pages 110–146)

1. Antoine Prost, *L'Enseignement s'est-il démocratisé?* (Paris, PUF, 1986).
2. Claude Thélot, *Tel père, tel fils? Position sociale et origine familiale* (Paris, Dunod, 1982), 144.
3. Jacques Lesourne, *Éducation et société. Les Défis de l'an 2000* (Paris, La Découverte, 1988), 211.
4. Prost, *L'Enseignement*, 133–9.
5. Pierre Bourdieu, *La Noblesse d'État. Grandes Écoles et esprit de corps* (Paris, Éditions de Minuit, 1989), 210.
6. Ezra N. Suleiman, *Elites in French Society. The Politics of Survival* (Princeton, NJ: Princeton UP, 1978), 87.
7. Thélot, *Tel père, tel fils?* 46, 101.
8. Merllié and Prévot, *La Mobilité sociale*, 62, 84.
9. Adeline Daumard, *Les Bourgeois et la bourgeoisie en France depuis 1815* (Paris, Flammarion, 1991), 357.
10. Institut Français d'Opinion Publique, *Le Climat politique en France au terme de la première législature* (Paris, 1962), 61.
11. Benoîte Groult, *Ainsi soit-elle* (Paris, Grasset & Fasquelle, 1976), 73.
12. Florence Montreynaud, *Le XXe siècle des femmes* (Paris, Nathan, 1992), 550.
13. Claire Duchen, *Feminism in France from May '68 to Mitterrand* (London, Routledge & Kegan Paul, 1986), 20.
14. Évelyne Sullerot, *Les Françaises au travail* (Paris, Hachette, 1973), 66–7.
15. Yvette Roudy, *La Femme en marge* (Paris, Flammarion, 1975), 11, 103–4.
16. Benoîte Groult, *Les Nouvelles Femmes* (Paris, Mazarine, 1979).
17. Monique Pelletier, *Nous sommes tous responsables* (Paris, Stock, 1981), 120.
18. *Libération*, 28 Oct. 1993.
19. Ibid. 22 Mar. 1994.
20. Françoise Giroud and Bernard-Henri Lévy, *Les Hommes et les femmes* (Paris, Olivier Orban, 1993), 20.
21. *Les Lettres françaises*, supplement to No. 18 (Mar. 1992), 'Où est passé le féminisme?'

22. Ernest Mignon, *Les Mots du Général* (Paris, Fayard, 1962), 57.

23. Commission de la Nationalité, *Être français, aujourd'hui et demain* (Paris, 1988), i. 445.

24. Gilles Kepel, *Les Banlieues d'Islam* (Paris, Seuil, 1987), 34.

25. Benjamin Stora, *La Gangrène et l'oubli. La Mémoire de la guerre d'Algérie* (Paris, La Découverte, 1991), 287.

26. *Le Choc du mois*, Jan. 1990, quoted in Christophe Bourseillier, *L'Extrême Droite* (Paris, Francis Bourin, 1991), 92.

27. Edmond Lipianski, *L'Identité française. Représentations, mythes, idéologies* (Paris, L'Espace Européen, 1991), 259.

28. *Le Monde*, 12 May 1993.

Chapter 6 (pages 147–168)

1. Tony Judt, *Past Imperfect. French Intellectuals, 1944–1956* (Berkeley, U. of California Press, 1992), 211.

2. Henri Lefebvre, *Everyday Life in the Modern World* (London: Allen Lane, The Penguin Press, 1971), title of ch. 2.

3. Pascal Dumontier, *Les Situationnistes et mai 68* (Paris, Gérard Lebovici, 1990).

4. *Noir et rouge. Anthologie 1956–1970* (Paris, 1982), 142–4.

5. *Le Monde*, 26 July 1983.

6. Alain Finkielkraut, *The Undoing of Thought* (London, Claridge Press, 1988), 116.

7. See above, p. 8.

8. Malraux, circular of 3 Feb. 1959, quoted in Jacques Charpentreau and René Kaës, *La Culture populaire en France* (Paris, Les Éditions Populaires/'Vivre son Temps', 1962), 101.

9. Jean-Jacques Lebel, *Procès du Festival d'Avignon, supermarché de la culture* (Paris, Pierre Belfond, 1968), 16.

10. Jean Baudrillard, *L'Effet Beaubourg. Implosion et dissuasion* (Paris, Éditions Galilée, 1977), 25.

Chapter 7 (pages 169–200)

1. Jacques Chaban-Delmas, *L'Ardeur* (Paris, Stock, 1975), 372.

2. Valéry Giscard d'Estaing, *Le Pouvoir et la vie* (Paris, Compagnie 12, 1988), 400.

3. Alain Bergounioux and Gérard Grunberg, *Le Long Remords du pouvoir. Le Parti socialiste français, 1905–1992* (Paris, Fayard, 1992), 326.

4. François Mitterrand, *La Rose au poing* (Paris, Flammarion, 1973), 28–9.

5. François Mitterrand, *Demain, Jaurès* (Paris, Pygmalion, 1977), 12.

6. *Le Monde*, 2 July 1981, cited in Olivier Duhamel, 'The Fifth Republic under François Mitterrand. Evolution and Perspectives', in Stanley Hoffmann (ed.), *The Mitterrand Experiment* (Cambridge, Polity Press, 1987), 143.

7. See pp. 98–9.
8. Pierre Favier and Michel Martin-Rolland, *La Décennie Mitterrand*, i: *Les Ruptures, 1981–1984* (Paris, Seuil, 1990), 418–19.
9. *Le Monde*, 16 Sept. 1983, cited in Jean-Marie Colombani and Jean-Yves Lhomeau, *Le Mariage blanc* (Paris, Grasset, 1986), 32.
10. Michel Rocard, *Le Cœur à l'ouvrage* (Paris, Seuil/Odile Jacob, 1987), 208.
11. François Furet and Pierre Rosanvallon, *La République du Centre* (Paris, Calmann-Lévy, 1988), 108.
12. *Le Monde*, 9 Apr. 1988.
13. See above, p. 136.
14. *Le Monde*, 16 Apr. 1994
15. Pascal Perrineau, 'Le Front National, 1972–92', in Winock, *Histoire de l'Extrême-Droite en France*, 293–4.
16. *Le Monde*, 4 Feb. 1994.
17. *Le Nouvel Observateur*, 20–6 Feb. 1992.
18. Philippe Boggio, *Coluche* (Paris, Flammarion, 1991), 242.

Chapter 8 (pages 201–226)

1. De Gaulle, *Discours et messages*, iii. 104–5.
2. Richard Shears and Isobelle Gidley, *The Rainbow Warrior Affair* (London, Counterpoint, 1986), 188.
3. *Le Monde*, 6 Mar. 1987, cited by Philippe Le Prestre (ed.), *French Security in a Disarming World* (London, Lynne Rieder, 1989), 26.
4. De Gaulle, *Discours et messages*, iv. 5
5. Alfred Grosser, *Affaires extérieures. La Politique de la France, 1944–1984* (Paris, Flammarion, 1984), 193.
6. Pierre-Bernard Cousté and François Visine, *Pompidou et l'Europe* (Paris, Librairies Techniques, 1974), 122.
7. *Le Monde*, 3 Sept. 1992.
8. Philippe Séguin, *Discours pour la France* (Paris, Grasset, 1992), 108.
9. *Le Monde*, 1 Sept. 1992.
10. Ibid., 4 Sept. 1992.
11. Ibid., 21 Sept. 1993.
12. Michel Aurillac, *L'Afrique à Cœur* (Paris, Berger-Levrault, 1987), 59.
13. *Le Monde*, 10 Nov. 1994.
14. Ibid., 6 May 1988.

Conclusion (pages 227–232)

1. *Libération*, 2 June 1993.
2. *Le Monde*, 20 Dec. 1994.
3. Ibid. 20 May 1989; 2 Sept. 1994.

Brief Chronology

1944	6 June	Allied landings in Normandy
	25 August	Liberation of Paris
	10 December	Signature of Franco-Soviet pact in Moscow
1945	31 March	French forces cross into Germany
	29 April/13 May	Municipal elections
	8 May	German capitulation; suppression of riots in Algeria
	29–31 May	Bombardment of Damascus
	2 September	Proclamation of independent Vietnamese Republic
	21 October	Constitutional referendum and elections to first Constituent Assembly
	13 November	De Gaulle elected head of provisional government
1946	20 January	Resignation of General de Gaulle
	23 January	Gouin government
	6 March	French recognition of Vietnamese Republic
	5 May	Referendum rejects first constitutional project
	28 May	Blum–Byrnes agreement
	1 June	Proclamation of Republic of Cochin-China
	2 June	Elections to second Constituent Assembly
	26 June	Bidault government
	13 October	Referendum endorses second constitutional project
	23 November	Bombardment of Haiphong
	16 December	Blum government
1947	16 January	Auriol elected president of the Republic
	28 January	Ramadier government
	February	Launch of Dior's New Look
	30 March	Insurrection in Madagascar
	7 April	Foundation of RPF
	4 May	Dismissal of Communist ministers
	17 June	France accepts Marshall Aid
	November	Wave of strikes and demonstrations
	22 November	Schuman government
1948	17 March	Signature of Brussels Pact
	27 July–27 August	Marie government

	11 September	Queuille government
1949	4 April	Signature of NATO Pact
	27 October	Bidault government
1950	3 March	Franco-German agreement on the Saar
	9 May	Schuman declaration on Coal and Steel Community
	13 July	Pleven government
	24 October	Pleven plan on European Army
1951	17 June	Legislative elections
1952	6 March	Pinay government
	27 May	Signature of EDC treaty
	28 May	Ridgway riots
1953	12 January	Opening of Oradour trial
	7 January–21 May	Mayer government
	26 June	Laniel government
	July	Launch of Poujadist movement
	23 December	Coty President of the Republic
1954	7 May	Fall of Dien Bien Phu
	18 June	Mendès France government
	? August	National Assembly rejects EDC
	1 November	Beginning of rebellion in Algeria
1955	6 February	Fall of Mendès-France government
	25 February	Faure government
	1–3 June	Messina conference
	23 November	Referendum in the Saar
1956	2 January	Legislative elections
	5 February	Mollet government
	4 November	Soviet repression in Hungary
	5–7 November	Franco-British raid on Suez Canal
1957	7 January	Full powers given to General Massu in Algiers
	25 March	Signature of Treaty of Rome
	21 May	Fall of Mollet government
	21 June–30 September	Bourgès-Maunoury government
	5 November	Gaillard government
1958	13 May	Pflimlin government; coup by extremists in Algiers
	1 June	Investiture of de Gaulle
	2 June	Full powers voted to de Gaulle
	14 September	Visit of Adenauer to de Gaulle
	28 September	Referendum on constitution of the Fifth Republic
	23/30 November	Legislative elections

	21 December	De Gaulle elected president of Fifth Republic and French Community
1959	9 January	Formation of Debré ministry
	March	Withdrawal of French Mediterranean fleet from control of NATO
	16 September	Speech of de Gaulle on self-determination of Algeria
1960	24 January–1 February	Week of the Barricades in Algiers
	13 February	Explosion of French atomic bomb in the Sahara
	January–July	Independence of French colonies in Sub-Saharan Africa
	5 September	Opening of Jeanson network trial; manifesto of 121 intellectuals on the Algerian cause
1961	8 January	Referendum on principle of Algerian self-determination
	22–5 April	Putsch of generals in Algiers
	17 October	Repression of Arab demonstration in Paris; scores of deaths
1962	8 February	Anti-OAS demonstration in Paris; eight deaths at Métro Charonne
	13 February	Demonstration for funeral of victims of Métro Charonne
	17 March	Evian agreement resulting in ceasefire in Algeria
	26 March	French army shoots at French Algerian demonstrators in Algiers
	8 April	Referendum on Algerian independence in metropolitan France
	14 April	Pompidou replaces Debré as prime minister
	1 July	Referendum of Algerian independence in Algeria
	22 August	Assassination attempt on de Gaulle at Petit-Clamart
	12 September	De Gaulle decides to hold referendum on direct elections to the presidency of the Republic
	5 October	Censure of Pompidou voted by National Assembly
	28 October	Referendum on direct elections to presidency of the Republic
	18/25 November	Legislative elections

1963	14 January	De Gaulle vetoes British application to join Common Market
	22 January	Franco-German treaty of co-operation
	22 June	Birth of French rock, Place de la Nation
1964	27 January	French recognition of People's Republic of China
	7 June	Creation of Convention des Institutions Républicaines
1965	1 July	French boycott of European Council of Ministers
	10 September	Creation of Fédération de la Gauche Démocrate et Socialiste
	5/19 December	Presidential elections; re-election of de Gaulle
1966	2 February	Foundation of Centre Démocrate
	4 March	France leaves integrated military command of NATO
1967	1 September	De Gaulle speech at Phnom Penh
	5/12 March	Legislative elections; setback for Gaullists
	5–10 June	Six-Day war
	26 July	De Gaulle speech at Montreal: 'Long live free Quebec!'
	19 December	Neuwirth law on contraception
1968	22 March	Student occupation at university of Nanterre
	3 May	Beginning of student unrest in Latin Quarter
	10–11 May	Night of the Barricades in Paris
	13 May	One-day strike; march of students and trade unions in Paris
	14–18 May	Official visit of de Gaulle to Romania
	27 May	Grenelle agreements with trade unions; rally in Charléty stadium
	29 May	Disappearance of de Gaulle
	30 May	Broadcast of de Gaulle; Gaullist rally in Paris
	23/30 June	Legislative elections; triumph of Gaullists
	10 July	Couve de Murville replaces Pompidou as prime minister
	21 August	Soviet invasion of Czechoslovakia
1969	27 April	De Gaulle loses referendum on reform
	28 April	Resignation of de Gaulle
	1/15 June	Presidential elections; Pompidou elected
	20 June	Chaban-Delmas appointed prime minister

	16 September	Chaban-Delmas speech on the New Society
1970	February	Visit of Pompidou to United States
1971	April	Manifesto demanding the right to abortion
	5 April	First screening of *Le Chagrin et la Pitié*
	June	Foundation of Socialist party at Épinay
1972	23 April	Referendum on British entry into Europe
	27 June	Socialist and Communist parties agree a Common Programme of Government
	5 July	Messmer replaces Chaban-Delmas as prime minister
1973	October	Beginning of oil crisis
	December	Solzhenitsyn's *Gulag Archipelago* published in Russian edition in Paris
1974	2 April	Death of Georges Pompidou
	5/19 May	Presidential elections; election of Giscard d'Estaing
	27 May	Jacques Chirac appointed prime minister
	November	PSU joins Socialist party
1975	17 January	Law on abortion promulgated
	November	Arms deal with Iraq
1976	February	Communist Party abandons doctrine of the dictatorship of the proletariat
	25 August	Resignation of Jacques Chirac; replaced as prime minister by Raymond Barre
	5 December	Foundation of RPR
1977	March	Municipal elections: success of the Left; election of Chirac mayor of Paris
	May	Foundation of Republican party
	May	Communist party endorses French nuclear deterrent
	14 September	Rupture of Union of the Left
	4 December	Coronation of Bokassa as emperor of Central African Republic
1978	January	Socialist party endorses French nuclear deterrent
	February	Foundation of the UDF
	12/19 March	Legislative elections; defeat of the Left
1979	13 March	France joins European Monetary System
	10 June	European elections
1980	19 May	Giscard d'Estaing meets Brezhnev in Warsaw

1981	26 April/10 May	Presidential elections; victory of François Mitterrand
	14/21 June	Legislative elections; PS and MRG win absolute majority
	22 June	Mauroy government includes four Communists
1982	2 March	Law on decentralisation
	March	Official visit of Mitterrand to Israel
	8 December	Resignation of Jean-Pierre Cot
1983	20 January	Mitterrand addresses Bundestag
	6/13 March	Municipal elections; setback for Socialists
	25 March	Austerity plan marks U-turn in Socialist economic policy
	13 July	Law on occupational equality
1984	January	Foundation of Greens as a united party
	17 June	European elections
	24 June	Demonstration by Catholics against school reform
	15 July	Resignation of Pierre Mauroy as prime minister; replaced by Laurent Fabius without Communists
	November	Launch of SOS-Racisme
1985	January	Jacques Delors takes up presidency of European Commission; idea of Single European Market launched
	10 July	Sinking of the *Rainbow Warrior*
1986	17–19 February	First summit of Francophone countries
	9/16 March	Legislative elections; victory for Right; Chirac appointed prime minister
	14 July	Row between Mitterrand and Chirac over privatization
	November–December	Student protests against university reform
1987	May–July	Trial of Klaus Barbie
1988	24 April/8 May	Presidential elections; reelection of Mitterrand
	10 May	Rocard replaces Chirac as prime minister
	14 May	Mauroy defeats Fabius for first secretaryship of Socialist party
	June	Legislative elections; return of Socialists with reduced majority
	November	Referendum on New Caledonia
1989	January	Pechiney affair discredits Socialist government
	June	European elections

	October	Muslim headscarf affair
	10 November	Fall of Berlin Wall; reunification of Germany begins
1990	March	Rennes congress of PS; Mauroy holds off challenge of Fabius for first secretaryship
	May	Foundation of Génération Écologie
1991	29 January	Resignation of Chevènement as foreign minister over Gulf War
	15 May	Rocard replaced as prime minister by Édith Cresson
	19 August	Mitterrand slow to condemn military coup against Gorbachev
	10 December	Treaty of Maastricht
1992	2 April	Cresson replaced as prime minister by Bérégovoy
	20 September	Referendum on Maastricht treaty
1993	March	Legislative elections; landslide of Right; Balladur prime minister
	1 May	Suicide of Pierre Bérégovoy
	13 May	Revision of law on French nationality
	October	Bourges congress of PS; Rocard secretary of the party
	19 November	Revision of constitution on right of asylum
	December	Conclusion of GATT talks
1994	16 January	Demonstration against reform of loi Falloux
	January	Resignation of Georges Marchais as secretary-general of PCF; PCF abandons doctrine of democratic centralism
	March	Demonstrations against 'SMIC-Jeunes'
	March–April	Trial of Paul Touvier
	12 June	European elections; setback for mainstream parties; resignation of Rocard as first secretary of PS
	September	Mitterrand affair Foundation of Independent Ecologist Movement
	December	Law against political corruption
1995	23 April/7 May	Presidential elections; election of Jacques Chirac

Further Reading

General

There are two excellent introductions to twentieth-century France: Maurice Larkin, *France Since the Popular Front. Government and People, 1936–1986* (Oxford, Clarendon Press, 1986) and James F. McMillan, *Twentieth-Century France: Politics and Society in France, 1898–1991* (London, Edward Arnold, 1992). In French there is René Rémond, *Notre siècle, 1918–1988* (Paris, Fayard, 1988). Julian Jackson's book on France between 1914 and 1992, to be published by Longman, is eagerly awaited. On the period since 1945 there are valuable translations of relevant volumes in the Seuil 'Points' series: Jean-Pierre Rioux, *The Fourth Republic, 1944–58* (Cambridge, CUP, 1987), and Serge Berstein, *The Republic of de Gaulle, 1958–1969* (Cambridge, CUP, 1993).

1. Crisis of Empire

General introductions to the issue of the status of France in the world include Alfred Grosser, *Affaires extérieures. La Politique de la France, 1944–1984* (Paris, Flammarion, 1984) and René Girault and Robert Frank (eds.), *La Puissance française en question, 1945–1949* (Paris, Publications de la Sorbonne, 1988). Charles de Gaulle, *Discours et messages*, i (1940–6), ii (1946–58), and iii (1958–62) (Paris, Plon, 1970) is essential. On Franco-American relations there are first-rate studies by Irwin W. Wall, *The United States and the Making of Postwar France, 1945–1954* (Cambridge, CUP, 1991) and Richard Kuisel, *Seducing the French: The Dilemma of Americanization* (Berkeley, U. of California Press, 1993), which may be supplemented by Denis Lacorne (ed.), *The Rise and Fall of Anti-Americanism. A Century of French Perception* (Basingstoke, Macmillan, 1990) and Frank Costigliola, *France and the United States. The Cold Alliance Since World War II* (New York, Twayne, 1992).

On France's relations with Germany and Europe in general, F. Roy Willis, *France, Germany and the New Europe, 1945–1967* (Oxford, OUP, 1967) still has mileage but may now be supported by John W. Young, *France, the Cold War and the Western Alliance, 1944–1949* (Leicester, Leicester UP, 1990), Klauss Manfrass and Jean-Pierre Rioux (eds.), *France–Allemagne, 1944–1947. Akten des deutsch–französischen Historiker Kolloquiums* (Baden–Baden, 2–5 Dec. 1986), Cahiers de l'Institut d'Histoire du Temps Présent, 13–14 (Dec. 1989, Jan. 1990), and Pierre Gerbet, *La Construction de l'Europe* (Paris, Imprimerie Nationale, 1983). Raymond Aron, *La Querelle de la CED* (Paris, Armand Colin, 1956) and Jean-Pierre Rioux, 'L'Opinion publique française et la Communauté Européenne de Défense', *Relations internationales*, 37 (1984), 37–53, cover the crisis over the EDC. Among important biographies or autobiographies of key participants in the building of Europe are Jean Monnet, *Memoirs* (London, Collins, 1978), Douglas Brinkley and Clifford Hackett, *Jean Monnet: The Path of European Unity* (Basingstoke, Macmillan, 1991), Raymond Poidevin, *Robert Schuman, homme d'état, 1886–1963* (Paris, Imprimerie Nationale, 1986), Robert Marjolin, *Architect of European Unity. Memoirs*, 1911–1986 (London, Weidenfeld & Nicolson, 1989), and Laurent Greilshamer, *Hubert Beuve-Méry, 1902–1989* (Paris, Fayard, 1990).

On the battle for France's colonies and decolonization in general, there is Raymond Betts, *France and Decolonisation*, 1900–1960 (Basingstoke, Macmillan, 1991) and Henri Grimal, *Decolonization. The British, French, Dutch and Belgian Examples* (London, Routledge & Kegan Paul, 1978). Grégoire Madjarian, *La Question coloniale et la politique du Parti Communiste Français, 1944–1947* (Paris, Maspéro, 1977) offers the Communist perspective. Syria and Indo-China are dealt with by A. B. Gaunson, *The Anglo-French Clash in Syria-Lebanon, 1940–1945* (Basingstoke, Macmillan, 1987), Jacques Dalloz, *The War in Indo-China, 1945–54* (Dublin, Gill & Macmillan, 1990), R. E. M. Irving, *The First Indo-China War. French and American Policy, 1945–54* (London, Croom Helm, 1975), and General Yves Gras, *Histoire de la guerre d'Indochine* (Paris, Denöel, 1992). To these may be added the comments of participants, such as General Henri Navarre, *Agonie d'Indochine, 1953–1954* (Paris, Plon, 1958), Joseph Laniel, *Le Drame indochinois. De Dien-Bien-Phu au pari de Genève* (Paris, Plon, 1957), and General Catroux, *Deux Actes du drame indochinois* (Paris, Plon, 1959). There are two collections on Pierre Mendès France: Janine Chêne, Edith Aberdam, and Henri Morsel (eds.), *Pierre Mendès France. La Morale en politique* (Grenoble, Presses Universitaires de Grenoble, 1990), and René Girault (ed.), *Pierre Mendès France et le Rôle de la France dans le monde* (Grenoble, Presses Universitaires de Grenoble, 1991). Older but still useful is Alexander Werth, *The Strange Story of Pierre Mendès France and the Great Conflict in French North Africa* (London, Barrie, 1957). The bibliography on the Algerian war is now enormous. Charles-Robert Ageron, *Modern Algeria. A History from 1830 to the Present* (London, Hurst, 1991), Alistair Horne, *A Savage War of Peace. Algeria 1954–1962* (Basingstoke, Macmillan, 1977), Patrick Eveno and Jean Planchais, *La Guerre d'Algérie* (Paris, La Découverte Le Monde, 1989), and Michael Kettle, *De Gaulle and Algeria, 1940–1960* (London, Quartet Books, 1993) are useful starting-points. On the impact of the war on various sections of the French military and public there are Jean-Pierre Vittori, *Nous, les appelés d'Algérie* (Paris, Stock, 1977), Martine Lemalet, *Lettres d'Algérie, 1954–1962. La Guerre des appelés. La Mémoire d'une génération* (Paris, Lattès, 1992), Patrick Rotman and Bertrand Tavernier, *La Guerre sans nom. Les Appelés d'Algérie, 1954–1962* (Paris, Seuil, 1992), Jean-Pierre Rioux and Jean-François Sirinelli, *La Guerre d'Algérie et les intellectuels français*, Cahiers de l'Institut d'Histoire du Temps Présent, 10 (Nov. 1988), David Schalk, *War and the Ivory Tower* (New York, OUP, 1991), Jean-Pierre Rioux (ed.), *La Guerre d'Algérie et les français* (Paris, Fayard, 1990), François

Bédarida and Étienne Fouilloux (eds.), *La Guerre d'Algérie et les chrétiens*, Cahiers de l'Institut du Temps Présent, 9 (Oct. 1988), and Daniele Joly, *The French Communist Party and the Algerian War* (Basingstoke, Macmillan, 1991). The issue of torture and opposition to the Algerian war is dealt with Henri Alleg, *La Question* (Paris, Pauvert, 1966), Maurice Maschino, *Le Refus* (Paris, Maspéro, 1960), Germaine Tillion, *Les Ennemis complémentaires* (Paris, Éditions de Minuit, 1960), Michel Auvray, *Objecteurs, insoumis, déserteurs. Histoire des réfractaires en France* (Paris, Stock, 1983), and Pierre Vidal-Naquet, *La Torture dans la République* (Paris, La Découverte Maspéro, 1983). The case for a French Algeria was made by Jacques Soustelle, *Le Drame algérien et la décadence française* (Paris, Plon, 1957) and *L'Espérance trahie, 1958–1961* (Paris, Alma, 1962), Pierre Lagaillarde, *On a triché avec l'honneur* (Paris, La Table Ronde, 1961), and Georges Bidault, *D'une Résistance à l'autre* (Paris, Presses du Siècle, 1965). The collective memory of the Algerian war is brilliantly analysed by Benjamin Stora, *La Gangrène et l'oubli. La Mémoire de la guerre d'Algérie* (Paris, La Découverte, 1992).

2. Crisis in the State

The best introductions to the politics of the Fourth Republic are Jean-Pierre Rioux, *The Fourth Republic, 1944–58*, cited above, and Georgette Elgey, *Histoire de la IVe République* (Paris, Fayard, 1965, 1968; new edn. 1992). On the alternatives facing France at the Liberation, Andrew Shennan, *Rethinking France. Plans for Renewal, 1940–1946* (Oxford, Clarendon Press, 1989), and François Bloch-Lainé and Jean Bouvier, *La France restaurée, 1944–1954* (Paris, Fayard, 1986) are essential. The question of a revolutionary situation in France is discussed by Grégoire Madjarian, *Conflits, pouvoirs et société à la Libération* (Paris, Union Générale d'Éditions, 1980), Annie Kriegel, *Communismes au miroir français* (Paris, Gallimard, 1974), Jean-Jacques Becker, *Le Parti Communiste veut-il prendre le pouvoir?* (Paris, Seuil, 1981), and Tony Judt (ed.), *Resistance and Revolution in Mediterranean Europe, 1939–1948* (London, Routledge, 1989). There are useful insights into the behaviour of de Gaulle in 1944–6 in Andrew Shennan, *De Gaulle* (London, Longman, 1993), Jean Lacouture, *De Gaulle* (New York, Norton, 1990), Claude Mauriac, *The Other de Gaulle. Diaries, 1944–1954* (London, Angus & Robertson, 1973), and Michel Debré, *Trois Républiques pour une France. Mémoires*, i, (Paris, Albin Michel, 1984). Vincent Auriol's *Journal du Septennat, 1947–1954* (7 vols., Paris, Armand Colin, 1970–8) is an indispensable reference for the workings of the Fourth Republic down to 1954. Among the political parties the Radicals are treated by Francis de Tarr, *The French Radical Party from Herriot to Mendès France* (London, OUP, 1961), Jean-Thomas Nordmann, *Histoire des radicaux, 1820–1973* (Paris, La Table Ronde, 1973), Serge Berstein, *Édouard Herriot ou la République en personne* (Paris, FNSP, 1985), and Pierre Delivet and Gilles Le Béguec (eds.), *Henri Queuille et la République. Actes du Colloque de Paris, Senate*, 25–6 Oct. 1984 (Paris, Trames, 1987). Works on the SFIO include Serge Berstein (ed.),

Paul Ramadier. La République et le socialisme (Paris, Complexe, 1990) and Bernard Ménager (ed.), *Guy Mollet, un camarade en République* (Lille, Presses Universitaires de Lille, 1987). The MRP is ill served, but Jean-Marie Mayeur, *Des Partis catholiques à la Démocratie Chrétienne, XIXᵉ–XXᵉ siècles* (Paris, Armand Colin, 1980) and Henri Descamps, *La Démocratie Chrétienne et le MRP de 1946 à 1959* (Paris, Librairie Générale de Droit et de Jurisprudence, 1981) lay the foundations. On the RPR, Jean Charlot, *Le Gaullisme d'opposition, 1946–1958* (Paris, Fayard, 1983) may be supplemented by Jacques Soustelle, *Vingt-huit ans de Gaullisme* (Paris, La Table Ronde, 1968), Jacques Chaban-Delmas, *L'Ardeur* (Paris, Stock, 1975), and the diaries of Claude Mauriac, cited above. Among the Independents, Pinay is the subject of biographies by Sylvie Guillaume, *Antoine Pinay ou la Confiance en politique* (Paris, FNSP, 1984) and by Christine Rimbaud, *Pinay* (Paris, Perrin, 1990), while Joseph Laniel has left his memoirs, *Jours de gloire et jours cruels, 1908–1958* (Paris, Presses de la Cité, 1971). On the Mendès France experiment, François Bédarida and Jean-Pierre Rioux (eds.), *Mendès France et le Mendésisme* (Paris, Fayard, 1985) may be supplemented by Pierre Mendès France, *Dire la vérité. Causeries du samedi, juin 1954–février 1955* (Paris, Julliard, 1955), his larger-scale *Œuvres complètes*, ii: *Une politique de l'économie, 1943–44*, iii: *Gouverner c'est choisir*, iv: *Pour une république moderne, 1955–62* (Paris, Gallimard, 1985–7), and Pierre Birnbaum, *Anti-Semitism in France. A Political History from Léon Blum to the Present* (Oxford, Blackwell, 1992). On Poujadism, Pierre Poujade, *J'ai choisi le combat* (Saint-Céré, Société Générale d'Éditions et des Publications, 1955), Stanley Hoffmann, *Le Mouvement Poujade* (Paris, Fondation Nationale des Sciences Politiques, 1956), and Dominique Borne, *Petits bourgeois en révolte? Le Mouvement Poujade* (Paris, Flammarion, 1977) suffice in advance of the new work by Richard Vinen.

On the last days of the Fourth Republic, to Michel Winock, *La République se meurt. Chronique, 1956–1958* (Paris, Seuil, 1978) may be added Françoise LeDouarec, *Félix Gaillard, 1919–1970. Un destin inachevé* (Paris, Economica, 1991) and Jean-Louis English and Daniel Rot, *Entretiens avec Pierre Pflimlin. Itinéraires d'un européen* (Strasburg, La Nuit Bleue, 1989).

Two recent biographies of de Gaulle, on different scales, are Andrew Shennan, *De Gaulle* (London, Longman, 1993) and Jean Lacouture, *De Gaulle* (New York, Norton, 1990). A recent account by one of his ministers is Alain Peyrefitte, *C'était de Gaulle* (Paris, Éditions de Fallois, 1994). Again, de Gaulle's *Discours et messages*, iii (1958–62) and iv (1962–9) (Paris, Plon, 1970) are an indispensable reference. Odile Rudelle, *Mai 58. De Gaulle et la République* (Paris, Plon, 1988) is an excellent analysis of the initial crisis, and can be read together with Michel Debré, *Trois Républiques pour la France. Mémoires, 1946–1958* (Paris, Albin Michel, 1988), Guy Mollet, *13 mai 1958–13 mai 1962* (Paris, Plon, 1962), and François Mitterrand, *Le Coup d'État permanent* (Paris, Plon, 1964). On the regime in general there is Serge Berstein, *The Republic of de Gaulle, 1958–1969* (Cambridge/Paris, CUP/Maison des Sciences de l'Homme, 1993) and, for a fairly hostile view, Pierre Viansson-Ponté, *Histoire de la*

République gaullienne (2 vols., Paris, Fayard, 1970–1). More specifically, the majority parties are dealt with by Jean Charlot, *L'UNR. Étude du pouvoir au sein d'un parti politique* (Paris, Armand Colin, 1967) and Jean-Claude Colliard, *Les Républicains Indépendants. Valéry Giscard d'Estaing* (Paris, PUF, 1971); and de Gaulle's prime ministers by the Institut Charles de Gaulle/Association Française de Science Politique, *De Gaulle et ses premiers ministres, 1959–1969* (Paris, Plon, 1990), Michel Debré, *Gouverner. Mémoires, iii: 1958–1962* (Paris, Albin Michel, 1988), and *Gouverner autrement. Mémoires, iv: 1962–1970* (Albin Michel, 1993), Gilles Martinet, *Le Système Pompidou* (Paris, Seuil, 1973), and Eric Roussel, *Georges Pompidou* (Paris, Lattès, 1984). Gaston Monnerville, *Vingt-deux ans de présidence* (Paris, Plon, 1980) is the critical perspective of the president of the Senate.

The Club movement is covered by Janine Mossuz, *Les Clubs et la politique en France* (Paris, Armand Colin, 1970), Philippe Reclus, *La République impatiente ou le Club des Jacobins, 1951–1958* (Paris, Publications de la Sorbonne, 1987), and Danièle Loschak, *La Convention des institutions républicaines. François Mitterrand et le socialisme* (Paris, PUF, 1971). Perspectives on the tribulations of the Left include François Mitterrand, *Ma part de vérité* (Paris, Fayard, 1969), Charles Hernu, *Priorité à gauche* (Paris, Denoël, 1969), Édouard Depreux, *Renouvellement du socialisme* (Paris, Calmann-Lévy, 1960), and *Souvenirs d'un Militant* (Paris, Fayard, 1972), Jean Poperen, *La Gauche française, i: Le Nouvel Âge, 1958–1965* (Paris, Fayard, 1972), Gaston Defferre, *Un Nouvel Horizon* (Paris, Gallimard, 1965) and *Si demain la Gauche* (Paris, Robert Laffont, 1977), and Christiane Hurtig, *De la SFIO au Nouveau Parti Socialiste* (Paris, Armand Colin, 1970).

The literature on May 1968 is voluminous. As a crisis of the regime it is best approached by Laurent Joffrin, *Mai 68. Histoire des événements* (Paris, Seuil, 1988), Jacques Capdevieille and René Mouriaux, *Mai 68. L'Entre-deux de la modernité. Histoire de trente ans* (Paris, FNSP, 1988), and from the memoirs of participants such as Michel Debré (see above), Georges Pompidou, *Pour rétablir la vérité* (Paris, Flammarion, 1982), and Alain de Boissieu, *Pour servir le Général, 1946–70* (Paris, Plon, 1982).

3. Echoes of the Occupation

The key text on this subject is Henry Rousso, *The Vichy Syndrome. History and Memory in France since 1944* (Cambridge, Mass., Harvard UP, 1991). To this has been added Eric Conan and Henry Rousso, *Vichy ou Les Dérives de la mémoire* (Paris, Fayard, 1994). Other studies of collective memories of the Occupation include Gérard Namer, *Mémoire et société* (Paris, Méridiens Klincksieck, 1987) and *La Commémoration en France de 1945 à nos jours* (Paris, L'Harmattan, 1987), Alfred Wahl (ed.), *Mémoire de la Seconde Guerre Mondiale. Actes du Colloque de Metz, 6–8 octobre 1983* (Metz, Centre de Recherche Histoire et Civilisation, 1984), Georges Kantin and Gilles Manceron (eds.), *Les Échos de la mémoire. Tabous et enseignement de la Seconde Guerre Mondiale*

(Paris, Le Monde Éditions, 1991). On the purges, the classic study is Peter Novick, *The Resistance versus Vichy. The Purge of Collaborators in Liberated France* (London, Chatto & Windus, 1968). To this may be added Marcel Baudot, 'La Résistance française face aux problèmes de répression et d'épuration', *Revue d'histoire de la Deuxième Guerre Mondiale*, 81 (Jan. 1971), 23–47, Herbert Lottman, *The People's Anger. Justice and Revenge in Post-Liberation France* (London, Hutchinson, 1986), and, for a hostile view, Philippe Bourdrel, *L'Épuration sauvage, 1944–1945* (Paris, Perrin, 1988). For the position of the extreme Right there are two seminal texts by Maurice Bardèche, *Lettre à François Mauriac* (Paris, La Pensée Libre, 1947) and *Nuremberg ou La Terre promise* (Paris, Les Sept Couleurs, 1948). The Resistance orthodoxy is set out by Henri Michel, *Histoire de la Résistance, 1940–1944* (Paris, PUF, 'Que sais-je?', 1950). The screenplay of Marcel Ophuls's film, *The Sorrow and the Pity*, is published by Paladin, 1972. The Jewish perspective on the Occupation and Resistance is elaborated in Serge Klarsfeld, *Le Mémorial de la déportation des juifs de France* (Paris, the author, 1978), Annie Kriegel, 'Résistants communistes et juifs persécutés', in her *Réflexions sur les questions juives* (Paris, Hachette, 1984), Annette Wieviorka, *Ils étaient juifs, résistants, communistes* (Paris, Denoël, 1986) and her *Déportation et génocide* (Paris, Plon, 1992). The revisionists are examined by Nadine Fresco, 'Les Redresseurs de torts', *Les Temps modernes*, 407 (June 1980), 2150–2211, and Pierre Vidal-Naquet, *Assassins of Memory* (New York, Columbia UP, 1992). On the Barbie trial there is Bernard-Henri Lévy, *Archives du procès Klaus Barbie* (Paris, Globe, 1986), Paul Gauthier (ed.), *Chronique du procès Barbie pour servir la mémoire* (Paris, Cerf, 1989)—a dossier of press comment—and Alain Finkielkraut, *La Mémoire vaine du crime contre l'humanité* (Paris, Gallimard, 1989). On Touvier, René Rémond (ed.), *Paul Touvier et l'Église. Rapport de la commission historique instituée par le Cardinal Decourtray* (Paris, Fayard, 1992), based on evidence provided by the Catholic Church, may be complemented by Arno Klarsfeld, *Touvier, un crime français* (Paris, Fayard, 1994). On Papon, there is Gérard Boulanger, *Maurice Papon. Un technocrate français sous l'Occupation* (Paris, Seuil, 1994). To conclude, Pierre Péan, *Une jeunesse française* (Paris, Fayard, 1994) unmasks the ambivalent relationship of François Mitterrand with the Resistance and Vichy.

4. Thirty Glorious, Twenty Inglorious Years.

Jean Fourastié's initial thesis, *Les Trente Glorieuses ou La Révolution invisible* (Paris, Fayard, 1979) has been supplemented by his further reflections (with Jacqueline Fourastié), *D'une France à une autre: avant et après les Trente Glorieuses* (Paris, Fayard, 1987).

Introductions to demographic, economic, and social developments include Henri Mendras with Alistair Cole, *Social Change in Modern France. Towards a Cultural Anthropology of the Fifth Republic* (Cambridge/Paris, CUP/Maison des Sciences de l'Homme, 1991), Daniel Noin and Yvan Chauviré, *La Population de la France* (Paris, Masson, 1987), Philip E. Ogden and Paul E. White, *Migrants in Modern France. Population Mobility in the Later Nineteenth and Twentieth Centuries* (London, Unwin Hyman, 1989), James F. Hollifield and George Ross, *Searching for the New France* (New York, Routledge, 1991), André Gueslin, *Nouvelle Histoire economique de la France contemporaine*, iv: *L'Économie ouverte, 1948–1990* (Paris, La Découverte, 1989), Maurice Parodi, *L'Économie et la société française depuis 1945* (Paris, Armand Colin, 1971), Jean-Marcel Jeanneney, *L'Économie française depuis 1967* (Paris, Seuil, 1989), Fernand Braudel and Ernest Labrousse (eds.), *Histoire économique et sociale de la France*, iv/3 (Paris, PUF, 1982), and Hubert Bonon, *L'Argent en France depuis 1880. Banquiers, financiers, épargnants dans la vie économique et politique* (Paris, Masson, 1989). On urbanization, there is Georges Duby (ed.), *Histoire de la France urbaine*, v: *La Ville aujourd'hui* (Paris, Seuil, 1985), Pierre Barrère and Micheline Cassou-Mounat, *Les Villes françaises* (Paris, Masson, 1980), Jean-Eudes Roullier, *Villes nouvelles en France* (Paris, Economica, 1989), Jacqueline Beaujeu-Garnier and Bernard Bézert, *La Grande Ville. Enjeu du XXIᵉ siècle* (Paris, PUF, 1991), and Jean-Paul Flamand, *Loger le peuple. Essai sur l'histoire du logement social* (Paris, La Découverte, 1989). On the planned economy, in addition to Andrew Shennan, *Rethinking France* and Bloch-Lainé and Bouvier, *La France restaurée* (see above), there are Richard Kuisel, *Capitalism and the State in Modern France. Renovation and Economic Management in the Twentieth Century* (Cambridge, CUP, 1981), Stephen Cohen, *Modern Capitalist Planning. The French Model* (Berkeley, of California Press, 1969), and Philippe Mioche, *Le Plan Monnet. Genèse et elaboration, 1941–1947* (Paris, Publications de la Sorbonne, 1987).

There is much good literature on agriculture and the peasant question, from Henri Mendras, *The Vanishing Peasant. Innovation and Change in French Agriculture* (Cambridge, Mass., MIT Press, 1970), Annie Moulin, *Peasant and Society in France Since 1789* (Cambridge/Paris, CUP/Maison des Sciences de l'Homme, 1991), and Gordon Wright, *Rural Revolution in France* (Stanford, Calif., Stanford UP, 1965), to Michel Gervais, Claude Servolin, and Jean Weil, *Une France sans paysans* (Paris, Seuil, 1965), Michel Debatisse, *La Révolution silencieuse. Le Combat des paysans* (Paris, Calmann-Lévy, 1963), a key text on the Jeunes Agriculteurs, Jean Chombart de Lauwe, *L'Aventure agricole en France de* 1945 à nos jours (Paris, PUF, 1979), Isabel Boussard, *Les Agriculteurs et la République* (Paris, Economica, 1990), Geneviève Gavignaud, *Les Campagnes en France au XXᵉ siècle, 1914–1989* (Paris-Gap, Ophrys, 1990) and Pierre Coulomb et al., *Les Agriculteurs et la politique* (Paris, FNSP, 1990).

On the working classes, the best introduction is Gérard Noiriel, *Workers in French Society in the 19th and 20th Centuries* (New York, Berg, 1990). Pierre Belleville, *Une nouvelle classe ouvrière* (Paris, Julliard, 1963) is really about the old working class; Serge Mallet, *La Nouvelle Classe ouvrière* (Paris, Seuil, 1963) is about the genuine article, as are Duncan Gallie, *In Search of the New Working Class. Automation and Social Integration within the Capitalist Enterprise* (Cambridge, CUP, 1979), which compares BP refineries in France and Great Britain, Nicole Eizner and Bertrand Hervieu, *Anciens Paysans, nouveaux ouvriers* (Paris, L'Harmattan, 1979), and Armand Frémont, *Ouvriers et Ouvrières à Caen* (Paris, CNRS, 1981). There are good monographs on individual industries or plants by Gérard Noiriel, *Longwy. Immigrés et prolétaires*, 1880–1980 (Paris, PUF, 1984), Pierre Naville *et al.*, *L'État entrepreneur. Le Cas de la régie Renault* (Paris, Anthropos, 1971), and Jacques Frémontier, *La Forteresse ouvrière: Renault* (Paris, Fayard, 1971). Good introductions to the middle classes are Christian Baudelot, Roger Establet, and Jacques Malemort, *La Petite Bourgeoisie en France* (Paris, Masson, 1974) and Adeline Daumard, *Les Bourgeois et la Bourgeoisie en France depuis* 1815 (Paris, Flammarion, 1991). On the *cadres* in particular Guy Groux, *Les Cadres* (Paris, La Découverte/Maspéro, 1983) provides an outline, Luc Boltanski, *The Making of a Class. Cadres in French History* (Cambridge/Paris, CUP/Maison des Sciences de l'Homme, 1987) a full study.

On the population, economy, and society since 1973, apart from Noin and Chauviré, Gueslin, Parodi, Jeanneney, Braudel and Labrousse, and Hollifield, already cited, there is John Gaffney (ed.), *France and Modernisation* (Aldershot, Avebury, 1988), John Tuppen, *France Under Recession,* 1981–1986 (Basingstoke, Macmillan, 1988), Pierre-Alain Muet and Alain Fonteneau, *Reflection and Authority. Economic Policy Under Mitterrand* (New York, Berg, 1990), the suggestive Élie Cohen, *L'État brancardier. Politiques du déclin industriel,* 1974–1984 (Paris, Calmann-Lévy, 1989), and the penetrating Dominique Taddéi (ed.), *'Made in France'. L'Industrie française dans la compétition mondiale* (Paris, Livre de Poche, 1993). *Le Monde* has produced two useful collections of data, Alain Gélédan (ed.), *Le Bilan économique des années Mitterrand,* 1981–1994 (Paris, Le Monde-Éditions, 1993) and *L'Économie française. Mutations,* 1975–1990 (Paris, Le Monde/Sirey, 1993). For the latest summaries there is the annual *L'État de la France* (Paris, La Découverte); the number for 1994–5 was published in 1994.

Studies of the decline of industry and the working class include Christian de Montlibert, *Crise économique et conflits sociaux dans la Lorraine sidérurgique* (Paris, L'Harmattan, 1989), Olivier Schwartz, *Le Monde privé des ouvriers. Hommes et femmes du Nord* (Paris, PUF, 1990), and Jean-Pierre Terrail, *Destins ouvriers. La Fin d'une classe?* (Paris, PUF, 1990). On changes in the countryside and the decline of the peasantry there is Roger Béteille, *La France du vide* (Paris, LITEC, 1981) and Pierre Alphandéry, Pierre Bitoun, and Yves Dupont, *Les Champs du départ. Une France rurale sans paysans?* (Paris, FNSP, 1990). On social fragmentation, welfare, and poverty there is Louis Roussel, *La Famille incertaine* (Paris, Odile Jacob, 1989), John S. Ambler, *The French Welfare State. Surviving Social and Ideological Change* (New York, New York UP, 1991), Jean Labbens, *Le Quart-Monde* (Pierrelaye, Éditions Science et Service, 1969), Claude Ferrand, *Exclusion et Sous-Prolétariat* (Paris, Programme, 1982), Louis Moreau de Bellaing, *La Misère blanche* (Paris, L'Harmattan, 1988), Serge Milano, *La Pauvreté absolue* (Paris, Hachette, 1988), Louise Camplong, *Pauvres en France* (Paris, Hatier, 1992), and Serge Paugam, *La Disqualification sociale. Essai sur la nouvelle pauvreté* (Paris, PUF, 1991).

5. The One and Indivisible Republic?

Optimistic views of the impact of education are put forward by Antoine Prost, *L'Enseignement s'est-il démocratisé?* (Paris, PUF, 1986) and Christian Baudelot and Roger Establet, *Le Niveau monte* (Paris, Seuil, 1989); more pessimistic views by Jacques Lesourne, *Éducation et société. Les Défis de l'an 2000* (Paris, La Découverte/Le Monde, 1988) and Philippe Raynaud and Paul Thibaud, *La Fin de l'école républicaine* (Paris, Calmann-Lévy, 1990). On social mobility there are classic studies by Claude Lévy-Leboyer, *L'Ambition professionnelle et la mobilité sociale* (Paris, PUF, 1971) and Raymond Boudon, *L'Inégalité des chances. La Mobilité sociale dans les sociétés industrielles* (Paris, Armand, Colin, 1973). The work of Claude Thélot, *Tel père, tel fils? Position sociale et origine familiale* (Paris, Dunod, 1982), which analyses social destinies in 1953 and 1977, has been incorporated into Yannick Lemel, *Stratification et mobilité sociale* (Paris, Armand Colin, 1991), continued for 1985 by Dominique Merllié and Jean Prévot, *La Mobilité sociale* (Paris, La Découverte, 1991). The outstanding work on the French élite is now Pierre Bourdieu, *La Noblesse d'État. Grandes Écoles et esprit de corps* (Paris, Éditions de Minuit, 1989), but more approachable are Jane Marceau, *Class and Status in France. Economic Change and Social Immobility, 1945–1975* (Oxford, OUP, 1977), Ezra N. Suleiman, *Elites in French Society. The Politics of Survival* (Princeton, NJ, Princeton UP, 1978), and essays by Jeanne Siwek- Pouydesseau and Daniel Derivry in Mattei Dogan (ed.), *The Mandarins of Western Europe. The Political Role of Top Civil Servants* (New York, Sage, 1975).

On the issue of women, Claire Duchen has now added *Women's Rights and Women's Lives in France, 1944–1968* (London, Routledge, 1994) to her *Feminism in France from May '68 to Mitterrand* (London, Routledge & Kegan Paul, 1986), which may be read alongside Françoise Picq, *Libération des femmes. Les Années mouvement* (Paris, Seuil, 1993). Two imaginative surveys are Florence Montreynaud, *Le XXᵉ Siècle des femmes* (Paris, Nathan, 1992) and Françoise Thiébaud's volume on the twentieth century in Georges Duby and Michelle Perrot (eds.), *A History of Women*, v (Cambridge, Mass., Belknap Press of Harvard University Press, 1992). Peter Morris (ed.), *Equality and Inequalities in France. Proceedings of the Fourth Annual Conference of the Association for the Study of Modern and Contemporary France* (1985) has an important section on women. Studies of women and work include Evelyn Sullerot and Françoise de Singly, *Fortune et infortune de la femme mariée* (Paris, PUF, 1987), while works on women and politics include Siân Reynolds, 'The French Ministry of Women's Rights, 1981–86: Modernisation or Marginalisation?', in John Gaffney (ed.), *France and Modernisation* (Aldershot, Avebury, 1988), 149–68, Choisir/La Cause des Femmes, *Fini le féminisme? Compte-rendu intégral du colloque international 'Féminisme et Socialismes', 13–15 Octobre 1983* (Paris,

Gallimard, 1984), and Françoise Gaspard, Claude Servan-Schreiber, and Anne Le Gall, *Au pouvoir citoyennes! Liberté, égalité, parité* (Paris, Seuil, 1992). Among key primary sources are Simone de Beauvoir, *The Second Sex* (London, David Campbell, 1953), Gisèle Halimi, *La Cause des femmes* (Paris, Grasset, 1973), Yvette Roudy, *La Femme en marge* (Paris, Flammarion, 1975), and *À cause d'elles* (Paris, Albin Michel, 1985), which relates to her ministerial experience, Benoîte Groult, *Ainsi soit-elle* (Paris, Grasset & Fasquelle, 1976) and *Les Nouvelles Femmes* (Paris, Marianne, 1979), Monique Pelletier, *Nous sommes toutes responsables* (Paris, Stock, 1981), the reflections of Giscard's secretary of state, and Françoise Giroud and Bernard-Henri Lévy, *Les Hommes et les femmes* (Paris, Olivier Orban, 1993).

On regional minorities there is Paul Sérant, *La France des minorités* (Paris, Robert Laffont, 1967). The changing discourse of regionalism may be explored in Robert Lafont, *La Révolution régionaliste* (Paris, Gallimard, 1967) and *L'Europe des ethnies* (Paris, Presses d'Europe, 1963). On the various cases of regionalism there is Michel Nicolas, *Le Séparatisme en Bretagne* (Brasparts, Éditions Bettan, 1986) and Maryon McDonald, *'We Are Not French!' Language, Culture and Identity in Brittany* (London, Routledge, 1989), Robert Lafont, *La Révolution occitane* (Paris, Flammarion, 1974), Pierre Letamienda, *Nationalismes au Pays Basque* (Bordeaux, Presses Universitaires de Bordeaux, 1987), and John Loughlin, *Regionalism and Ethnic Nationalism in France. A Case Study of Corsica* (Florence, European University Institute, 1989). There is now an authoritative study of decentralization in Vivien A. Schmidt, *Democratizing France. The Political and Administrative History of Decentralization* (Cambridge, CUP, 1990).

On the various religions flourishing in France are Guy Michelat, Julien Potel, Jacques Sutter, and Jacques Maître, *Les Français sont-ils encore catholiques?* (Paris, Cerf, 1991), Gérard Cholvy and Yves-Marie Hilaire, *Histoire religieuse de la France contemporaine*, iii: 1930–88 (Toulouse, Privat, 1988), Dominique Schnapper, *Jewish Identities in France. An Analysis of Contemporary French Jewry* (Chicago, U. of Chicago Press, 1983), Doris Bensimon and Sergio della Pergola, *La Population juive en France: Socio-démographie et identité* (Paris, Institute of Contemporary Jewry/Hebrew University of Jerusalem/CNRS, 1986), Judith Friedlander, *Vilna on the Seine: Jewish Intellectuals in France Since 1968* (New Haven, Conn., Yale UP, 1990), Frank Ezkenazi and Édouard Waintrop, *Le Talmud et la République* (Paris, Grasset, 1991), Gilles Kepel's masterly *Les Banlieues de l'Islam* (Paris, Seuil, 1987), Rémy Leveau and Gilles Kepel, *Les Musulmans dans la société française* (Paris, FNSP, 1988), and Bruno Étienne, *La France et l'Islam* (Paris, Hachette, 1989). On immigration, Olivier Milza, *Les Français devant l'immigration* (Paris, Complexe, 1988) and François Dubet, *Immigrations: qu'en savons-nous? Un bilan de connaissances* (Paris, La Documentation Française, 1989) are useful introductions, but Yves Lequin (ed.), *La Mosaïque France. Histoire des étrangers et de l'immigration* (Paris, Larousse, 1988) and Gérard Noiriel, *Le Creuset français. Histoires de l'immigra-*

tion, XIXe–XXe siècles (Paris, Seuil, 1988) are indispensable. Questions of assimilation, integration, and their relationship to French national identity are tackled by Dominique Schnapper's outstanding *La France de l'intégration. Sociologie de la nation* (Paris, Gallimard, 1991), Étienne Balibar and Immanuel Wallerstein, *Race, Nation, Class. Ambiguous Identities* (London, Verso, 1991), Pierre-André Taguieff 's penetrating *Face au racisme*, i: *Les Moyens d'agir; ii: Analyses, hypothèses, perspectives* (Paris, La Découverte, 1991, 1992), Edmond Lipianski, *L'Identité française. Représentations, mythes, idéologies* (Paris, L'Espace Européen, 1991), and Suzanne Citron, *Le Mythe national. L'Histoire de France en question* (Paris, Les Éditions Ouvrières, 1989). North Africans in particular are examined by Jean-Mars Terrasse, *Génération Beur* (Paris, Plon, 1986) and Christian Jelen, *Ils feront de bons français. Enquête sur l'assimilation des maghrébiens* (Paris, Robert Laffont, 1991), while the issue of immigrants and political rights is dealt with by Catherine Wihtol de Wenden, *Les Immigrés et la politique* (Paris, FNSP, 1988) and Jacqueline Costa-Lascoux, *De l'immigré au citoyen* (Paris, La Documentation Française, 1989). The proceedings of the Commission de la Nationalité, *Être français aujourd'hui et demain* (2 vols., Paris, 1988), contains many interesting testimonies. Finally, work on racism includes Pierre-André Taguieff, *La Force du préjugé. Essai sur le racisme et ses doubles* (Paris, La Découverte, 1988) and Michel Wieviorka, *La France raciste* (Paris, Seuil, 1992).

6. Cultural Revolutions

On intellectuals in general Louis Bodin, *Les Intellectuels* (Paris, PUF, 'Que sais-je?', 1962) is still a good starting-point. Régis Debray, *Teachers, Writers, Celebrities. The Intellectuals of Modern France* (London, New Left Books/Verso, 1981), Pascal Ory and Jean-François Sirinelli, *Les Intellectuels en France de l'Affaire Dreyfus à nos jours* (Paris, Armand Colin, 1986), and Sirinelli, *Intellectuels et passions françaises. Manifestes et pétitions au XXe siècle* (Paris, Fayard, 1990) are essential, while Jeremy Jennings, *Intellectuals in Twentieth-Century France. Mandarins and Samurais* (Basingstoke, Macmillan, 1993) contains some useful essays. On Sartre, existentialism, communism, and the intellectual controversies of post-war France there is Annie Cohen-Solal, *Sartre. A Life* (London, Heinemann, 1987), Anna Boschetti, *Sartre et les 'temps modernes'* (Paris, Éditions de Minuit, 1985), Herbert Lottman, *The Left Bank. Writers, Artists and Politics from the Popular Front to the Cold War* (London, Heinemann, 1987), Arthur Hirsch, *The French New Left. An Intellectual History from Sartre to Gorz* (Boston, South End Press, 1981), Jeannine Verdès-Leroux, *Au service du Parti. Le Parti Communiste, les intellectuels et la culture, 1944–1956* (Paris, Fayard/Minuit, 1983), Ariane Chebel d'Appollonia, *Histoire politique des intellectuels en France, 1944–1954* (Paris, Complexe, 1991), Tony Judt, *Past Imperfect. French Intellectuals, 1944–1956* (Berkeley, U. of California Press, 1992), Sunil Khilnani, *Arguing Revolution. The Intellectual Left in Postwar France* (New Haven, Conn., Yale University Press, 1993). On varieties of Marxism and 1968, Hirsch and Khilnani may be supplemented by Henri Lefebvre, *Everyday Life in the Modern World* (London, Allen Lane, 1971), Guy Debord, *The Society of the Spectacle* (Detroit/New York, Black and Red/Zone Books, 1970), Jeannine Verdès-Leroux, *Le Réveil des somnambules. Le Parti Communiste, les intellectuels et la culture, 1956–1965* (Paris, Fayard/Minuit, 1987), Pascal Dumontier, *Les Situationnistes et mai 68* (Paris, Gérard Lebovici, 1990), Luc Ferry and Alain Renaut, *La Pensée 68. Essai sur l'anti-humanisme contemporain* (Paris, Gallimard, 1985), and François Bédarida and Michel Pollack (eds.), *Mai 68 et les sciences sociales*, Cahiers de l'Institut d'Histoire du Temps Présent, 11 (Apr. 1989), and Keith Reader, *Intellectuals and the Left in France Since 1968* (Basingstoke, Macmillan, 1987). John Sturrock (ed.), *Structuralism and Since* (Oxford, OUP, 1979) is a good introduction to structuralist thought; on individual thinkers there is Jonathan Culler, *Barthes* (London, Fontana, 1983), Edmund Leach, *Lévi-Strauss* (London, Fontana/Collins, 1970), Malcolm Bowie, *Lacan* (Cambridge, Mass., Harvard UP, 1991), Sherry Turkle, *Psychoanalytic Politics. Freud's French Revolution* (New York, Basic Books, 1978), J. G. Merquior, *Foucault* (London, Fontana, 1985), Mark Poster, *Foucault, Marxism and History* (Cambridge, Polity Press, 1984), and Christopher Norris, *Derrida* (London, Fontana, 1987). The revival of right-wing thought may be followed in Hirsch, *The French New Left*, Khilnani, *Arguing Revolution*, André Glucksman, *La Cuisinière et le mangeur d'hommes. Essai sur l'état, le marxisme, les camps de concentration* (Paris, Seuil, 1975), Bernard-Henri Lévy, *La Barbarie au visage humain* (Paris, Grasset & Fasquelle, 1977) and *L'Idéologie française* (Paris, Grasset, 1981), and Anne-Marie Duranton-Crabol, *Visages de la nouvelle droite. La GRECE et son histoire* (Paris, FNSP, 1988). Suggestive on the decline of the French intellectual are Maurice Blanchot, 'Les Intellectuels en question', *Le Débat*, 29 (Mar. 1984), Jean-François Lyotard, 'Tombeau de l'intellectuel', in *Tombeau de l'intellectuel et autres papiers* (Paris, Galilée, 1984), and Alain Finkielkraut, *The Undoing of Thought* (London, Claridge Press, 1988).

On mass culture in general, the first-rate survey of cultural practices, *Les Pratiques culturelles des français, 1973–1989,* by the Département d'Études et de la Prospective/Ministère de la Culture et de la Communication (Paris, La Découverte/La Documentation Française, 1990) should be supplemented by Pascal Ory, *L'Entre-Deux-Mai. Histoire culturelle de la France, mai 1968–mai 1981* (Paris, Seuil, 1983), his rather better *L'Aventure culturelle française, 1945–1985* (Paris, Gallimard, 1985), Joffre Dumazedier, *Révolution culturelle du temps libre, 1968–1988* (Paris, Méridiens-Klincksieck, 1988), and Brian Rigby and Nicholas Hewitt (eds.), *France and the Mass Media* (Basingstoke, Macmillan, 1991). Different branches of the mass media are dealt with by Jean-Louis Missika and Dominique Wolton, *La Folle du logis. La Télévision dans les sociétés démocratiques* (Paris, Gallimard, 1983), Jacques Durand, *Le Cinéma et son public* (Paris, Sirey, 1958), Pierre Sorlin, *European Cinemas, European Societies, 1939–1990* (London, Routledge, 1991), Antoine Virenque, *L'Industrie cinématographique française* (Paris, PUF, 'Que sais-je?', 1990), and Pierre Albert, *La Presse française* (Paris, La Documentation Française, 1990).

On cultural policies since the war there are good studies by Evelyne Ritaine, *Les Stratèges de la culture* (Paris, FNSP, 1983) and Pierre Cabane, *Le Pouvoir culturelle sous la Ve République* (Paris, Olivier Orban, 1981). Augustin Girard and Geneviève Gentil, *Cultural Developments, Experiences and Policies* (2nd edn., Paris, Unesco, 1983) provides useful theoretical insights. Jack Lang's system is analysed by David Loosely, 'Jack Lang and the Politics of Festival', *French Cultural Studies*, i/i (1990), 5–19, and attacked by Guy Hocquenghem, *Lettre ouverte à ceux qui sont passés du col de Mao au Rotary* (Paris, Albin Michel, 1986) and especially by Marc Fumaroli, *L'État culturel* (Paris, Éditions de Fallois, 1992).

7. The Republic of the Centre

The concept of the Republic of the Centre comes from François Furet, Jacques Julliard, and Pierre Rosanvallon, *La République du centre. La Fin de l'exception française* (Paris, Calmann-Lévy, 1988). On political parties and the electorate in general there are Alistair Cole, *French Political Parties in Transition* (Aldershot, Dartmouth, 1990), John Frears, *Parties and Voters in France* (London, Hurst, 1991), Frédéric Le Bon and Jean-Paul Cheylan, *La France qui vote* (Paris, Hachette, 1988), and Colette Ysmal, *Le Comportement électoral des français* (Paris, La Découverte, 1990). On the Pompidou era, Gilles Martinet, *Le Système Pompidou* (Paris, Seuil, 1973) and Eric Roussel, *Pompidou* (Paris, Lattès, 1984) may be supplemented by Jean Bunel and Paul Meunier, *Chaban-Delmas* (Paris, Stock, 1972), Philippe Alexandre, *Exécution d'un homme politique* (Paris, Grasset, 1973), on the toppling of Chaban, and Jacques Chaban-Delmas, *L'Ardeur* (Paris, Stock, 1975). The face-to-face debates of the 1974 presidential elections are recorded and analysed in Valéry Giscard d'Estaing and François Mitterrand, *54774 mots pour convaincre* (Paris, PUF, 1976). On the Giscard presidency, John R. Frears's rather thin *France in the Giscard Presidency* (London, Allen & Unwin, 1981) should be supplemented by Jean-Christian Petitfils, *La Démocratie giscardienne* (Paris, PUF, 1981). Giscard's own writings, *Démocratie française* (Paris, Fayard, 1976) and *Le Pouvoir et la vie* (Paris, Compagnie 12, 1988), are worth reading. On his prime ministers there is Franz-Olivier Giesbert, *Jacques Chirac* (Paris, Seuil, 1987) and Raymond Barre, *Réflexions pour demain* (Paris, Hachette, 1984).

The trials of the parties of the Left are dealt with by Annie Kriegel, *Un autre communisme* (Paris, Hachette, 1977), Jean-Jacques Becker, *Le Parti Communiste veut-il prendre le pouvoir?* (Paris, Seuil, 1981), Olga Narkiewicz, *The End of the Bolshevik Dream. Western European Communist Parties in the Late Twentieth Century* (London, Routledge, 1990), Sudhir Hazareesingh, *Intellectuals and the French Communist Party. Disillusion and Decline* (Oxford, Clarendon Press, 1991), Marc Lazar, *Maisons rouges. Les Partis Communistes français et italien de la Libération à nos jours* (Paris, Aubier, 1992), David Bell and B. Criddle,

The French Communist Party in the Fifth Republic (Oxford, Clarendon Press, 1994), David S. Bell and B. Criddle, *The French Socialist Party. The Emergence of a Party of Government* (2nd edn., Oxford, Clarendon Press, 1988), Alain Bergounioux and Gérard Grunberg, *Le Long Remords du pouvoir. Le Parti Socialiste français, 1905–1992* (Paris, Fayard, 1992). Among studies of Mitterrand are Catherine Nay, *Les Sept Mitterrand, ou Les Métamorphoses d'un Septennat* (Paris, Grasset, 1988), Eric Roussel, *Mitterrand, ou La Constance du Funambule* (Paris, Lattès, 1991), Wayne Northcutt, *Mitterrand. A Political Biography* (New York, Holmes & Meier, 1992), and Alistair Cole, *François Mitterrand. A Study in Political Leadership* (London, Routledge, 1994). Mitterrand's own writings include *Ma part de Vérité* (Paris, Fayard, 1969), evidently not the whole truth, *Le Socialisme du possible* (Paris, Seuil, 1971), on the need for a break with capitalism, *La Rose au poing* (Paris, Flammarion, 1973) and *L'Abeille et l'architecte. Chronique* (Paris, Flammarion, 1978). Other reflections on socialism include Michel Rocard, *À l'épreuve des faits. Textes politiques, 1979–1985* (Paris, Seuil, 1986) and *Le Cœur à l'Ouvrage* (Paris, Seuil, 1987), and Laurent Fabius, *C'est en allant vers la mer* (Paris, Seuil, 1990). On the Socialists in power the literature includes Pierre Birnbaum (ed.), *Les Élites socialistes au pouvoir, 1981–1985* (Paris, PUF, 1985), Stanley Hoffmann (ed.), *The Mitterrand Experiment* (Cambridge, Polity Press, 1986), Thomas R. Christofferson, *The French Socialists in Power, 1981–1986. From Autogestion to Cohabitation* (Newark, NJ, U. of Delaware Press/London, Associated University Press, 1991), Solange and Christian Gras, *Histoire de la première république mitterrandienne* (Paris, Robert Laffont, 1991), Pierre Favier and Michel Martin-Rolland, *La Décennie Mitterrand*, i: *Les Ruptures, 1981–1984*, ii: *Les Épreuves* (Paris, Seuil, 1990–1), and Eric Dupin, *L'Après Mitterrand. Le Parti Socialiste à la dérive* (Paris, Calmann-Lévy, 1991), which deals with Fabius, Jospin, and Rocard. The experience of cohabitation between 1986 and 1988 is examined by Jean-Marie Colombani and Jean-Yves Lhomeau, *Le Mariage blanc* (Paris, Grasset, 1986), Maurice Duverger, *La Cohabitation des français* (Paris, PUF, 1987), John Tuppen, *Chirac's France, 1986–1988* (Basingstoke, Macmillan, 1991), and Édouard Balladur, *Passion et longeur de temps. Dialogues avec Jean-Pierre Elkabbach* (Paris, Fayard, 1989). On the presidential elections of 1988 there is John Gaffney (ed.), *The French Presidential Elections of 1988* (Aldershot, Dartmouth, 1989), on the relationship between Mitterrand and Rocard, Robert Schneider, *La Haine tranquille* (Paris, Seuil, 1992), and the last Socialist premier, Christiane Rimbaud, *Bérégovoy* (Paris, Perrin, 1994). On the Balladur era there is Stanley Hoffmann, 'Keeping Demons at Bay', *New York Review of Books*, 3 Mar. 1994, Philippe Bauchard, *Deux ministres tranquilles* (Paris, Belfond, 1994), on Balladur and Bérégovoy, and Edwy Plenel, *Un temps de chien* (Paris, Stock, 1994), on Balladur and Mitterrand.

There is now a considerable literature of high quality on the Front National. This includes Edwy Plenel, *L'Effet Le Pen* (Paris, La Découverte, 1984), Nonna Meyer and Pascal Perrineau (eds.), *Le Front National à découvert* (Paris, FNSP,

1989), Michel Winock, *Nationalisme, anti-sémitisme et fascisme en France* (Paris, Seuil, 1990), Birgitta Orfali, *L'Adhésion au Front National* (Paris, Kimé, 1990), Christophe Bourseillier, *L'Extrême Droite. L'Enquête* (Paris, François Bourrin, 1991), Madeleine Rebérioux, *L'Extrême Droite en questions* (Paris, Études et Documentation Internationales, 1991), Guy Birnbaum, *Le Front National en politique* (Paris, Balland, 1992), Nonna Meyer and Pascal Perrineau, 'Why Do They Vote for Le Pen?', *European Journal of Political Research*, 22 (1992), 123–41, Edwy Plenel and Alain Rollat, *La République menacée. Dix ans d'Effet Le Pen* (Paris, Le Monde-Éditions, 1992), and Pascal Perrineau, 'Le Front National, 1972–1992', in Michel Winock (ed.), *Histoire de l'Extrême Droite en France* (Paris, Seuil, 1993). A starting-point for Le Pen's own rhetoric is his *Les Français d'abord* (Paris, Carrère/Michel Laffon, 1984). Writings on the ecology movement and Green party are less voluminous but include Alain Touraine, *Anti-Nuclear Protest. The Opposition to Nuclear Energy in France* (Cambridge/Paris, CUP Maison des Sciences de l'Homme, 1983), Guillaume Sainteny, *Les Verts* (Paris, PUF, 'Que sais-je?', 1991), Christain Brodhag, *Objectif Terre. Les Verts, de l'écologie à la politique* (Paris, Éditions du Félin, 1990), which deals mainly with ecologist thought, Brendan Prendiville and Tony Chafer, 'Activists and Ideas in the Green Movement in France', in Wolfgang Rüdig (ed.), *Green Politics One* (Edinburgh University Press, 1990), and Brendan Prendiville, *L'Écologie. La Politique autrement?* (Paris, L'Harmattan, 1993). Works on political corruption include Gilles Gaetner, *L'Argent facile. Dictionnaire de la corruption en France* (Paris, Stock, 1992) and Yves Mény, *La Corruption dans la République* (Paris, Fayard, 1992). Approaches to alternative politics include Philippe Boggio, *Coluche* (Paris, Flammarion, 1991), Harlem Désir, *Touche pas à mon pote* (Paris, Grasset, 1985), and Serge Malik, *Histoire secrète du SOS Racisme* (Paris, Albin Michel, 1990).

8. France in Search of a World Role

General surveys include Alfred Grosser, *Affaires extérieures. La Politique de la France, 1944–1984* (Paris, Flammarion, 1984), Robert Aldrich and John Connell, *France in World Politics* (London, Routledge, 1989), and Françoise de La Serre *et al.* (eds.), *French and British Foreign Policy in Transition. The Challenge of Adjustment* (New York, Berg, 1990).

Foreign policy is the first concern of presidents of the Republic. Their views and studies of their foreign policy may be found in Charles de Gaulle, *Discours et messages*, iii (1958–62), iv (1962–64), v (1964–69) (Paris, Plon, 1970), Philip G. Cerny, *The Politics of Grandeur. Ideological Aspects of de Gaulle's Foreign Policy* (Cambridge, CUP, 1980), Georges Pompidou, *Entretiens et discours, 1968–1974* (2 vols., Paris, Plon, 1975), Valéry Giscard d'Estaing, *L'État de la France* (Paris, Fayard, 1981) and *Le Pouvoir et la vie* (Paris, Compagnie 12, 1988), Samy Cohen and Marie-Claude Smouts, *La Politique extérieure de Valéry Giscard d'Estaing* (Paris, FNSP, 1985), François Mitterrand, *Réflexions sur la politique extérieure de la France* (Paris, Fayard, 1988), Stanley Hoffmann,

'Mitterrand's Foreign Policy or Gaullism By Any Other Name', in Hoffmann (ed.), *The Mitterrand Experiment* (Cambridge, Polity Press, 1987).

The question of the relationship of France to the superpowers is explored in Denis Lacorne, *The Rise and Fall of Anti-Americanism. A Century of French Perception* (Basingstoke, Macmillan, 1990), Frank Costigliola, *France and the United States. The Cold Alliance since World War II* (New York, Twayne, 1992), Richard Kuisel, *Seducing the French. The Dilemma of Americanization* (Berkeley, U. of California Press, 1993), and Robert O. Paxton (ed.), *De Gaulle and the United States. A Centennial Reappraisal* (Oxford, Berg, 1994). To these may be added Michael M. Harrison, *The Reluctant Ally. France and Atlantic Security* (Baltimore, Johns Hopkins UP, 1981), Jean-Jacques Servan-Schreiber's classic *The American Challenge* (London, Hamish Hamilton, 1968), André Wilmots, *Le Défi français, ou La France vue par l'Amérique* (Paris, François Bourin, 1991), Dominique Moisi, 'Franco-Soviet Relations and French Foreign Policy', in Paul Godt (ed.), *Policy-making in France from de Gaulle to Mitterrand* (London, Pinter, 1989).

The debate on the nuclear deterrent is examined in Diana Johnstone, 'How the French Left Learned to Love the Bomb', *New Left Review*, 46 (July–Aug. 1984), 5–36, Jolyon Howorth and Patricia Chilton (eds.), *Defence and Dissent in Contemporary France* (London, Croom Helm, 1984), Richard Shears and Isobelle Gidley, *The Rainbow Warrior Affair* (London, Counterpoint, 1986), Philippe Le Prestre, *French Security Policy in a Disarming World* (Boulder, Colo., Lynne Rieder, 1989), and Jean d'Albion, *Une France sans défense* (Paris, Calmann-Lévy, 1991).

Works on the relationship of France with Europe include Roger Massip, *De Gaulle et l'Europe* (Paris, Flammarion, 1963), Karl W. Deutsch et al. (eds.), *France, Germany and the Western Alliance. A Study of Elite Attitudes on European Integration and World Politics* (New York, Scribner, 1966), Pierre Maillard, *De Gaulle et l'Allemagne. Le Rêve inachevé* (Paris, Plon, 1990), and Pierre-Bernard Cousté and François Visine, *Pompidou et l'Europe* (Paris, Librairies Techniques, 1974). Debates on European union and German unification may be followed in Raoul Girardet (ed.), *La Défense de l'Europe* (Paris, Complexe 1988), an exchange between party politicians in Institut de Relations Internationales et Stratégiques, *La Défense de la France dans les années 90* (Paris, La Documentation Française, 1990), Laurent Cohen-Tanugi's wide-ranging *L'Europe en danger* (Paris, Fayard, 1992), Jacques Delors and Clisthène, *Our Europe* (London, Verso, 1992), and Dave Berry and Martyn Cornick, 'French Responses to German Unification', *Modern and Contemporary France*, 49 (Apr. 1992), 42–55.

On France's relationship with her dependencies, former colonies, and the Third World, Jacques Adda and Marie-Claude Smouts, *La France face au Sud. Le Miroir brisé* (Paris, Karthala, 1989) is excellent, Régis Debray, *Tous Azimuts* (Paris, Seuil, 1989) rather polemical. On black Africa there are John Chipman, *French Power in Africa* (Oxford, Blackwell, 1989) and Francis Terry McNamara, *France in Black Africa* (Washington, DC, National Defense UP, 1989). Two very different accounts by Ministers of Co-operation are Jean-Pierre Cot, *A l'épreuve du pouvoir. Le Tiers-mondisme, pourquoi faire?* (Paris, Seuil, 1984) and Michel Aurillac, *L'Afrique à Cœur* (Paris, Berger-Levrault, 1987). Robert Aldrich and John Connell, *France's Overseas Frontier. Départememts et Territoires d'Outre-Mer* (Cambridge, CUP, 1992) is authoritative on the DOM-TOMs. Guides to Francophonie include Xavier Deniau, *La Francophonie* (Paris, PUF, 'Que sais-je?', 1983), William Bostock, *Francophonie. Organisation, Coordination, Evaluation* (Fitzroy, Victoria, Australia, Seine Publications, 1986), and Haut Conseil de la Francophonie, *État de la francophonie dans le monde* (Paris, La Documentation Française, 1987).

Index

Note: names of political parties, with the exception of the Communist Party and Radical Party, are given in French for greater precision.